Charles Evans Reeves

Heart Diseases in Australia

With observations on aneurism of the aorta

Charles Evans Reeves

Heart Diseases in Australia
With observations on aneurism of the aorta

ISBN/EAN: 9783337315603

Printed in Europe, USA, Canada, Australia, Japan

Cover: Foto ©berggeist007 / pixelio.de

More available books at **www.hansebooks.com**

HEART DISEASES

IN

AUSTRALIA,

WITH OBSERVATIONS ON

ANEURISM OF THE AORTA.

BY

C. E. REEVES, B.A., M.D.,

AUTHOR OF "DISEASES OF THE STOMACH AND DUODENRUM;" "DISEASES OF THE SPINAL CORD AND MEMBRANES;" "HYDROCEPHALUS;" "SOFTENING OF THE STOMACH IN AUSTRALIA;" AND "ON A SIMPLE MODE OF TREATING STRICTURE," &C., &C.

"Nulla est alia pro certo noscendi via nisi quam plurimas et morborum et dissectionem historias, tum aliorum, tum proprias, collectas habere, et inter se comparare."

Morgagni. De Sed, et Causis, Morb. lib. IV. proam.

PUBLISHED BY THE AUTHOR,

AND SOLD BY

J. BROOKS, 238 BOURKE STREET EAST, MELBOURNE.

1 8 7 3.

TO

The Memory

MY DEAR AND LAMENTED FRIEND,

CORNELIUS STEWART, M.D.

CONTENTS.

CHAPTER I.

HEART DISEASES IN AUSTRALIA.

CHAPTER II.

CHOREA OF THE HEART.

CHAPTER III.

NERVOUS PALPITATION OF THE HEART.

CHAPTER IV.

INFLAMMATION OF THE PERICARDIUM—*(PERICARDITIS).*

CHAPTER V.

DISEASE OF THE VALVES OF THE HEART.—INFLAMMA-
TION OF THE LINING MEMBRANE OF THE HEART.—
(*ENDOCARDITIS.*)

CHAPTER VI.

DISEASES OF THE RIGHT SIDE OF THE HEART.—
(*ENDOCARDITIS.*)

DISEASES OF THE PULMONARY ARTERY AND VALVES.

CHAPTER VII.

HYPERTROPHY OF THE HEART.—ENLARGEMENT OF
THE HEART.

CHAPTER VIII.

FATTY DISEASE OF HEART AND FLABBY HEART.

CHAPTER IX.

1. RUPTURE OF THE HEART. 2. RUPTURE OF THE VALVES. 3. CANCER. 4. HYDATIDS. 5. ANGINA PECTORIS.

CHAPTER X.

ANEURISM OF THE THORACIC AORTA.

INTRODUCTION.

THE following pages contain the results of a more or less careful examination of between five and six hundred cases of Disease of the Heart.

The tables have been formed from the cases which presented themselves at the Institution for Diseases of the Chest, established seven years ago. In addition to this, a large number of cases were collected in the Melbourne Hospital, in private practice, from cases which either came under my own or the care of my friends.

The diseases incidental to the heart in the colony do not present any very marked differences from those observed in Europe in the symptoms observed during life, and the changes found after death, beyond being of a less inflammatory type. Diseases of a nervous character are more common, and the same may be said of hypertrophy, dilatation, and fatty degeneration.

The chapters on Chorea of the Heart—an affection nearly unknown in cold climates; on Nervous Palpitation, so common amongst females here; Diseases of the Pulmonary Artery and the right side of the Heart—Hypertrophy, Fatty Disease, &c.—will be found to contain some observations which may not be without interest to the profession in Europe. The chapter on Aneurism of the Aorta, a disease which is exceedingly frequent here, has been made as exhaustive as possible, without giving a history of every case observed.

I would particularly wish to draw the attention of the profession to the temperature of the breath in health, after sleep, food, and stimulants, and in disease. It is a field in which no one has hitherto laboured, and promises to be of far greater value than the examination of the temperature of the surface of the body. In exhausting diseases it will be found to sink gradually; before death, when fluid exists in the bronchial tubes, it will sink below the temperature of the air; while in fever, inflammation, after stimulants, and in hot weather, it will rise. I use a piece of perforated cedar or pine of the shape of a respirator, with a small thermometer inserted so that the breath can pass freely around it during expiration; inspiration being performed by the nose. By means of a small mirror, the patient can note the temperature at different times of the day and under different circumstances.

I am sensible that I expose myself to the charge of presumption in attempting to elucidate such a complicated class of diseases as those of the heart. If some of the subjects are not so fully illustrated as the European student would have an opportunity of doing, the defect must be attributed to the absurd restrictions placed on the permission granted to medical men to visit our hospital to study the diseases incidental to the colony.

Whether the object of an institution intended not only for the relief of suffering humanity, but for the advancement of medical science, and supported by large grants from the public purse, is carried out by the present exclusive system, is doubtful. I do not think that I err in stating that one of the noblest institutions in these colonies is looked upon by a clique as their special property. How long this kind of thing is to exist, is is impossible to say— perhaps until some scarifactor shall spring up and lay it open.

It is, no doubt, out of place to attempt to publish a book on any disease of the colony, when men who can scarcely put twelve words together daily advertise themselves as authors of works, paraphrased from some American edition of a German or French author.* In Europe, these quasi authors would be quickly exposed, but here these literary robberies pass unchallenged. Many hundred years ago, Martial applied the lash to those who made fame and profit on others' labour, in something like these words :—

> " Quâ factus ratione sit, requiris,
> Qui nunquam futuit, pater Jacobus ?
> H ——————— dicat, istud,
> *Qui scribit nihil, tamen auctor est.*" †

and then salved the wheals with—

> " Carmina Jacobus emit : recitat sua carmina Jacobus."
> " Nam, quod emas, possis dicere jure tuum."

This may be translated—

> " If James does emit his little song,
> What odds? he's bought and paid for it ;
> Let James alone—to emit on."

---

* There has been for some time a paraphrase of Hempel's translations of Jahr's work on syphilis, with " appropriate treatment and plates," very freely scattered about Melbourne. The titles of the *author* (3), formed into a pyramid of the size of that of Cheops, form the most amusing illustration of charlatanic fanfaronade in the English language.

† Butler *(Satire on Plagiaries)* must have had these lines in his mind when exposing Sir John Denham's purchase of the poem of Cooper's Hill, "Writ by a vicar who had forty pound for it." He says :—

> " ——————— They strut and swell
> Like thieves bedecked with what they steal.
> *   *   *   *   *   *   *   *
> Literary fathers, who, like the cuckold and sot,
> Are proud of what they ne'er begot."

Less fortunate men can only quote the great Roman's Satire—

> " Is there no bridge vacant, no convenient seat,
> Where thou may'st cringe, and gnaw thy rotting meat,
> And with a stick for crutch, and mock-ulcerated leg,
> More *honestly* and *honourably* beg."

C. E. REEVES, M.D.

Collins-street, Melbourne, June, 1873.

Heart Diseases in Australia.

CHAPTER I.

General Observations.—Frequency of Heart Diseases in the Colony, compared with England.—Causes of Heart Diseases in the Colony.—Relative Frequency of the different Forms.—Influence of the Seasons in causing them.—Mortality in Victoria and in the different Hospitals.—Deaths in England.

Diseases of the Heart are nearly as fatal in the colony as in England. They are perhaps more frequent, particularly the nervous and fatty forms, and they generally occur from the same causes—rheumatism and over-exertion. The most common causes of the nervous forms are disorders of the functions of the womb and ovary—over-suckling—excessive use of spirits or tobacco. In young lads and females, abuse of the sexual organs is a great cause of this form of the disease.

Diseases of the stomach and liver are often a source of palpitation. Distension of the stomach may, by pressing up the liver, interfere with the action of the heart; and enlargement of the left lobe of the liver will act in a similar way. The left arch of the colon, if loaded with fæces, tumors, and fluid, into the cavity of the abdomen, may also, by pressing up the liver, produce all the symptoms of disease, and even an alteration in the valvular sounds.

There is another cause of the nervous form observed here amongst females at the change of life, from the 35th to the 50th year. It seems to depend as much upon the presence of some poison in the blood as on nervous exhaustion. The symptoms are very common, and bear a very close resemblance to those observed in colonial fever. It is observed in those females who are employed in housework, and who often pass from a hot room into a cold place or damp yard. The poison generated by the perspiration being checked does not seem to be able to produce an attack of colonial fever, as it would do in a young or unacclimatised person—but it causes what females generally call a succession of " hot flushes," which may affect the face or the whole body, leaving the skin covered with a clammy tasteless or saltish sweat, which is very rarely acid, unless rheumatism exists. The sweat has a peculiar smell, and the urine, which will be found to vary in quantity and colour and the amount of phosphates it contains, is nearly certain to be as offensive as putrid fish or meat. The skin is dry and harsh, or greasy—the patient becomes thin—she is easily fatigued— is nervous and irritable, and the thirst marked, but hot fluids encourage the hot flushes. The palpitation of the heart at first comes on, sometimes with the "hot flushes" irregularly, but particularly on the slightest excitement. In the course of time, if the patient is obliged to continue working, disease of the valves, with or without hypertrophy of the walls of the heart, may set in.

There is considerable difficulty in tracing these cases from their onset to their termination in disease of the heart. The poison generated in colonial fever is capable of producing inflammation of the lining membrane of the blood vessels and of the heart, but not so rapidly, or so frequently, as that of rheumatism. Now and then cases of colonial fever will be met with when the patient has slept for some time in a cold, damp room, in which the valves of the heart become affected. Congestion of the lungs, particularly of the posterior lobes, is a nearly constant attendant on colonial fever, lying on the back favoring its occurrence.—' In all cases in which this congestion is due to poison in the blood, there is a tendency, which cannot be however always easily traced, to implication of the lining membrane of the right side of the heart, and even of the left side and of the aorta. This implication of the right side can generally be most easily seen when the pulmonary veins are very much congested.

It is often extremely difficult to distinguish between organic disease of the heart and sympathetic disorder. For instance, it is not uncommon, particularly in disease of the mitral and pulmonary valves, to find the patients liable to attacks of palpitation, with more or less difficulty of breathing, after the first sleep, or early in the morning. In an elderly woman, who had repeatedly suffered from attacks of nervous palpitation, which were relieved by iron and quinine, and who at last became subject to mitral diseases enlargement of the liver, and dropsy—the attacks of palpitation and difficulty of breathing, just at daybreak, were very troublesome, usually lasting an hour. When they ceased, she could lie down and walk about without difficulty.

The sound over the mitral valve, in front and behind, along the border of the scapula, was rather harsh, particularly in the former situation. During the attacks of difficulty of breathing, the heart's action was greatly increased, and the sound in the mitral valve was very much intensified, as if spasmodic contraction of the opening existed. This alteration in the mitral and pulmonary valves will be observed in an affection peculiar to the heart, and is of common occurrence in the colony—namely, chorea of the heart, an affection unknown in Europe.

These observations will tend to show how necessary it is to carefully distinguish between sympathetic and organic diseases of the heart. The palpitation depending on disease of the stomach—liver—womb—or exhaustion of the nervous system, or anæmia, is sure to be aggravated by the exhibition of digitalis—veratrium viridium, or other depressants of its action, and if too freely given they may cause syncope, and even death.

The following table, showing the relative frequency of various diseases of the heart, may not be without interest :—

Of 310 cases treated, by far the greatest number—127—were suffering from nervous disease of the organ.

Simple hypertrophy, and hypertrophy with alterations in the aortic and mitral valves, were the next most frequent—54 out of the 310 cases.

Alterations in the aortic—mitral and pulmonary valves, next—44—being about as frequent as pericarditis—45.

On analysing the 310 cases—

102 were cases of palpitation without hypertrophy, or alteration in the aortic or mitral valves.

25 were cases of chorea.

29 were cases of simple hypertrophy, without alteration of the mitral or aortic valves.

25 were cases of hypertrophy, with alteration of the mitral and aortic valves, in some cases of only one, but in most cases of both.

6 were cases of alteration of the aortic valves, without hypertrophy.

14 were cases of alteration of the mitral valves.

18   „   „   of the aortic and mitral valves.

6 were cases of alteration of the pulmonary valves.

13 ,, ,, of fatty degeneration.

7 ,, ,, of flaccid heart.

3 ,, ,, of angina pectoris.

45 ,, ,, of pericarditis—either acute or chronic. In 20 there was alteration either of the aortic or mitral valves; sometimes of both.

17 were cases of fibrinous clots in the heart.

310 Total.

In the cases in which fibrinous clots formed in the ventricles and large vessels, they occurred after confinement in two cases; in hæmorrhage from the lungs in eight; in diarrhœa or dysentery, anæmia and mental anxiety, with chronic disease of the heart or lungs, in seven. There were only three cases of angina pectoris seen. The patients were old men, their ages being 60, 70, and 73 years.

Influence of the Seasons in causing Diseases of the Heart.

From rheumatism being more liable to occur in cold weather, pericarditis and disease of the valves are more frequently observed than in mild dry weather. In cold weather, following on hot days, pleuritis and congestion of the lungs seem to be most commonly met with. This was the case during the months of September of 1871 and 1872. In several of the cases of pleuritis, the inflammation extended from the pleura to the pericardium. In pleuro-pneumonia of the left side of the chest, it is not uncommon to find a considerable quantity of serum in the pericardium; the pericardium may be even found more or less adherent. It is not so common to find, as in Europe, sero-purulent fluid, with flakes of lymph in the pericardium, or layers of fibrinous deposits on thè heart. When these are met with, the patients have been either very much exposed to cold, or they have re-resided in a colder district than that in which Melbourne is situated. It will, I think, be generally observed that the colder the season the more the diseases will be found to resemble in type those observed in England.

In winter here, the type of the diseases met with is more in-flammatory, the exudation more fibrinous, than in the hot weather, when serous effusion is more liable to occur.

Old cases of disease of the heart are apt to end fatally in the very hot weather, from congestion of the lungs—pulmonary apoplexy—or from the action of the skin or kidneys being interfered with from exposure to the cold south wind, which will often cause the temperature to sink suddenly from 110, or even 117 or 120 deg., 30 or 40 or even 50 degrees, chilling the patients through as severely as if they were plunged into ice.

To those suffering from heart disease or consumption, if reduced in strength, these sudden changes of temperature are very injurious.

The winter of 1870 was very cold and wet. Rheumatism was exceedingly common, and diseases of the heart were therefore very frequent, both with it and subsequently from the patients being exposed too soon to cold. This after implication was particularly common among delicate boys and girls.

They usually made a good recovery from the rheumatism, but from exposure symptoms of heart disease would set in, the mitral valve being the most liable to be affected. The winters of 1871 and 1872 having been mild, rheumatism and heart diseases were much less frequent.

The mortality in Victoria from rheumatism—pericarditis—

diseases of the heart, and aneurism, as I before observed, is very large, as will be seen from the following table :—

In 1868, the deaths were, from

Rheumatism	...	50.— 33	males.	17	females.
Pericarditis	...	22.— 15	„	7	„
Heart diseases	...	372.—244	„	128	„
Aneurism		62.— 57	„	5	„
Total	...	349	„	157	„

Rheumatism has been included in the table, from death, when it does prove fatal, being generally from disease of the heart.

The following table has been compiled from the Melbourne Hospital Reports, from 1857 to 1869, inclusive. It is not of any value beyond showing the number of admissions and number of deaths from rheumatism. In 1861, in less than twelve months, I noted 11 cases of pericarditis in the wards, but there is not one case returned in that year.

The number of admissions in that year from rheumatism was large, being 256, and the deaths 5 :—

	No. of Cases Treated.	Cases of Aneurism.	Heart Disease.	Pericarditis.	Rheumatism.
1857...	1683	8—3 deaths.	56—23 deaths.	...	135—11 deaths.
1858...	2013	3—2 „	55—16 „	4—2 deaths.	113—4 „
1859...	2967	1 (Disease of artery)	47— 8 „	...	227—6 „
1860...	3628	17 (Diseases of arteries)	66—14 „	...	216—6 „
1861...	4309	1	60—13 „	...	256—5 „
1862...	3179	9—3 deaths. 10 (Arteries— Diseases of) —9 deaths.	29—11 „	...	209—1 „
1863...	3147	20—2 „	67—35 „	5—2 „	158—2 „
1864...	3024	14—7 „	57—34 „	2	177—1 „
1865...	3253	19—9 „	38—10 „	6—3 „	302—6 „
1866...	3300	16—6 „	70—35 „	6—1 „	297—4 „
1867...	3095	16—6 „	123—43 „	...	185—7 „
1868...	3255	14—10 „	99—38 „	...	237—8 „
1869...	3355	26—8 „	99—46 „	2—1 „	252—7 „

The following table, compiled from the Registrar-General's returns, may be of interest to the profession. It will show the number of admissions and deaths in the different Colonial Hospitals, in 1868, from rheumatism—pericarditis—heart diseases —and all causes :—

	Rheumatism.		Pericarditis.		Heart Disease		All Causes.	
	Admissions.	Deaths.	Admissions.	Deaths.	Admissions.	Deaths.	Admissions.	Deaths.
Amherst.............	25	...	...	...	13	2	217	24
Ararat...............	30	...	1	...	10	3	285	26
Ballarat	81	1	2	1	33	8	1090	113
Beechworth	12	...	...	...	9	4	392	50
Belfast	3	...	...	...	...	...	44	6
Bendigo	60	2	4	3	20	9	734	89
Castlemaine	46	...	4	1	5	1	620	46
Creswick...........	11	...	1	1	5	1	238	17
Daylesford*	6	...	...	...	4	2	150	13
Dunolly	13	1	...	...	4	1	244	28
Geelong*	105	...	...	...	15	4	958	64
Hamilton*	21	...	...	...	1	...	210	17
Heathcote	6	...	1	...	7	4	92	11
Inglewood	8	...	...	...	6	2	123	12
Kilmore	11	...	...	...	2	1	125	9
Kyneton............	25	...	1	...	8	3	289	19
Maldon	3	...	...	...	1	1	91	12
Maryborough......	29	1	...	...	4	1	274	20
Melbourne	237	8	...	...	99	38	3255	423
Pleasant Creek ...	39	...	2	...	8	3	360	27
Portland*	1	...	...	...	...	...	27	2
Sale.................	12	...	2	...	7	2	136	12
Swan Hill	12	...	...	...	3	...	108	3
Warrnambool ...	5	...	...	...	4	...	91	5
Wood's Point......	11	...	...	...	1	...	99	4

In the colony, the deaths from rheumatism and diseases of the heart are much greater among males than females than in England. This is due to men being much more exposed to the weather than females. The mortality from aneurism is also much greater among males than females. This is the reverse of what is ob-

* These Hospitals are also Benevolent Asylums.

served in England; the number of deaths, with the exception of aneurism, which is more fatal among males than females, being nearly equal, as will be seen from the following table :—

Deaths in England, in 1862, from

Rheumatism	... ...	984	males.	959 females.
Heart diseases ...	...	8685	,,	9692 ,,
Pericarditis	... ...	270	,,	289 ,,
Aneurism	... ...	294	,,	79 ,,
Rheumatism, with pericarditis, or disease of the heart	...	349	,,	355 ,,
Carditis	... ...	45	,,	68 ,,
Endocarditis	... ...	35	,,	45 ,,

In London, in the same year, there died from

Rheumatism	... ...	196	males.	190 females.
Heart diseases ...	...	1359	,,	1481 ,,
Pericarditis	... ...	53 .	,,	58 ,,
Aneurism	... ...	92	,,	11 ,,

The estimated population of England was 20,336,467.

In Monmouthshire and North and South Wales, with a population of 1,212,834, the deaths from

Rheumatism, were	...	55	males.	54 females.
Heart diseases ...	...	390	,,	382 ,,
Pericarditis	... ...	12	,,	12 ,,
Aneurism	... ...	5	,,	2 ,,
Total	... ...	462	,,	450 ,,

From this it will be seen that in this district, with about double the population of Victoria, 905 died from rheumatism—pericarditis, and heart disease, while in Victoria 444 died of the same diseases.

CHAPTER II.

CHOREA OF THE HEART.

CHARACTERS.—FREQUENCY IN THE COLONY.—COMMON AMONGST YOUNG MEN AND WOMEN.—ITS CONNECTION WITH CHOREA OF THE LIMBS, &C., AND WITH DISEASES OF THE VALVES OF THE HEART.—CASES.—DISEASE OF THE HEART AND CHOREA.—DIAGNOSIS,—CAUSES.—TREATMENT.

CHARACTERISTIC FEATURES.—*Violent attacks of palpitation of variable duration, occurring when the patients are perfectly quiet—generally at night, during the first sleep—compelling them, in the severer form, to jump out of bed, and even to rush into the open air.—After the attacks pass off, the patients, although weak, are able, if no organic disease exists, to follow their usual occupations, and walk fast without any inconvenience.*

In Europe this affection of the heart is but very rarely observed, but here it is very common. It occurs in the young of both sexes, and when uncomplicated with chorea, or disease of the heart, generally arises from self-abuse. The excessive use of tobacco, drinking spirits in large quantities, and late hours, seem to have much to do with favoring its occurrence.

Chlorotic females are very liable to it. In these cases, with the disordered menstrual discharge, leucorrhœa is nearly certain to exist.

In both sexes it is not uncommon to find that there is some choreic affection of some of the muscles of the body.

The action of the heart will be generally found increased, and often accompanied with some alteration in the valvular sounds in chorea of the limbs.

When chorea occurs in those suffering from disease of the valves of the heart, the sounds become louder, as if the orifice was more contracted than at other times. The mitral and pulmonary seem to be more susceptible than the aortic and the right auriculo-ventricular valves.

In disease of the mitral and pulmonary valves, there is often a tendency to periodical attacks of palpitation, and which, like the uncomplicated form of chorea of the heart, may occur in the night.

The following are the cases in which the relation between the action of the heart and the choreic movements were well marked :—

In the first case the patient, aged 23 years, was a trooper in the police. He was thin, but muscular, and had enjoyed good health until 18 months ago. Then, without any previous notice, he was awoke from his first sleep with violent palpitation and difficulty of breathing. He rushed into the open air, and the attack passed off in three or four minutes. He then returned to bed, but the attack returned. He again jumped up, but this time he fell, but did not lose consciousness. Now the attacks occur every night; they vary in intensity, sometimes being severe, but at others they occur as shocks of palpitation just as he is about to fall asleep, and cause him to jump up in bed. Of late he has become liable to attacks of faintness in the daytime, with palpitation. These attacks occur at irregular times, without any warning. They are so severe that he is in dread of dying suddenly in one of them. His pulse is feeble, ranging from 90 to 96. In the interval of the attacks, he is able to do his duty—walk, and even run with ease.

Mental excitement—anxiety of mind, and smoking to excess—increase the heart's action, and cause the night attacks to be severer than they otherwise would be. The only alteration about the heart was a blowing sound heard along the pulmonary artery. This sound I found was greatly increased in intensity during the attacks of palpitation. The pulse then rose to 120, became full and hard, and the face flushed. His manner was restless—he kept constantly shifting his position in the chair, drawing his legs under it, and then as suddenly extending them. This led me to suspect the nature of the disease. On inquiry, he said that he suffered from irregular attacks of jerkings of the arms, legs, or head. In walking, his legs gave a peculiar twitch just as he was about to raise his foot. The affection of the heart and the jerking of the limbs had been very much aggravated by strychnine, which had been given him two or three weeks before. Twenty minims of tincture of opium, with ten of digitalis, checked the choreic symptoms in a few minutes, and caused the blowing sound to disappear. He had recourse to the same draught on two other occasions, and each time the attack was cut short. Small doses of arsenic, with sulphate of zinc, digitalis, and morphia were ordered three times a day, with a belladonna and opium plaster to the chest.

It is scarcely necessary to trace the progress of this case from the time he was first seen, on the 20th of October, up to the 20th of December, when the blowing sound heard over the valves of the pulmonary artery had quite disappeared. It had gradually diminished in intensity. On the 16th of November he had only slight jerking of the hands at times; the sound was then very feeble. On the 27th, he had had "no jerks or palpitation for some time;" the sound was then so feeble that it could be scarcely heard. He had several relapses while in the force, but when last seen he had not had an attack for eighteen months.

In the case of a young man, a clerk, aged 19, suffering from permanent alteration in the pulmonary valves, there was a tendency, during mental excitement, to chorea of the muscles of the left side of the neck, face, and arm. The heart's action became at the same increased—the alteration in the pulmonary valves louder, and there was then slight alteration heard over the mitral valve. In this case, opium aggravated the symptoms.

Alteration was heard over the mitral valve in several other cases in which chorea existed with palpitation of the heart. In one case, the patient, a female, 17 years of age, had previously had an attack of chorea, which lasted five or six months. She had never been quite free—for over-exertion and mental anxiety still caused slight jerkings of the right arm to occur at night. From a fright, the jerking had become as severe as ever, and she was unable to lift a cup of tea to her mouth with the right arm. Her menstrual discharge was regular and natural in color, but she suffered from leucorrhœa after its cessation. It was at this time that the jerkings of the limbs were most severe. The palpitation of the heart also became troublesome, particularly at night, when it awoke her, and caused her to sit up in bed. The attacks lasted from five to ten, and even fifteen minutes. If she went to bed very much fatigued, or had been excited during the day, the attacks were generally worse. There was a distinct blowing sound heard over the mitral valve. Her pulse was natural—face somewhat pale, but not markedly chlorotic. She had never had rheumatism.

In the case of a female, aged 11 years, who had been suffering for six weeks from chorea of the arms and left leg, there was an alteration in the heart's sounds. As the chorea disappeared, the sounds became natural.

In another case, the alteration in the first sounds of the heart was more marked when the chorea, which affected the left half of the body, was severe. Several times the chorea disappeared, but the alteration in the heart sounds continued. When it disappeared, there was no return of the chorea.

The connection between disease of the heart and chorea has not received that attention from medical writers which it seems to demand. In this climate, from over-educating and working young females, and shutting them up in hot, close rooms, both chorea and chlorosis are becoming frequent.

With the increase of these diseases, the connection between chorea and diseases of the heart will be more fully elucidated than in Europe. Willan, as early as 1801, published two cases in which chorea existed in connection with disease of the heart. In both cases, fluid existed in the pericardium and in the ventricles of the brain.

Ten years later, Forgues published a case in which lymph and serum existed in the pericardium ; and, in 1818, Dr. Copland one in which, with the fluid in the pericardium, there was hypertrophy of the left ventricle. In 1824, Abercrombie observed a case in which the pericardium was dark-colored, vacular, and adherent, and covered with lymph. Prichard, about the same time, found the pericardium adherent in a case. Roeser, Richard Bright,

Tong, Babington, Addison, Todd, Burrows, Favel, Taylor, Hughes, Peacock, Gaff, Lee, Nairne, Omerod, Kirkes, and others, have published similar cases.

In 1858, I collected a number of cases, with the view of drawing the attention of the profession to the connection of chorea and disease of the heart.

Thirty-four cases were collected ; in 5 of this number the disease of the heart occurred unconnected with rheumatism. Symptoms of heart disease had existed for some time before the chorea set in. In one case, suppression of the menstrual discharge had excited chorea once before ; the patient had suffered from disease of the heart—a fresh attack of heart disease from exposure to cold, and in the course of a few days the chorea appeared. Its appearance was preceded by great agitation, nervousness, and loss of sleep, and with disturbance of the heart's action. In one of the 23 cases in which the chorea occurred in connection with rheumatism and disease of the heart, on the appearance of the chorea the abnormal sounds subsided, but the heart continued to act tumultuously.

DIAGNOSIS.—This affection is nearly always considered to be organic disease, and is, therefore, treated as such. Digitalis, veratium viridium, and medicines of a similar character, are given, but with no other effect than to aggravate the symptoms, and if too freely given, to produce fatal syncope. It bears a very strong resemblance to angina pectoris, and to the symptoms present in disease of the valves. Chorea is a disease which occurs early in life, while angina is rarely observed, even in this colony, where deposits form in the coats of blood-vessels much earlier than in England—before the 45th year. In disease of the pulmonary valves—although there is an intimate connection between the two diseases—the sound is permanent in the valves, and the difficulty of breathing, with disturbance of the heart's action, on ascending a hill or walking fast is always the same. (*See Diseases of the Mitral and Pulmonary Valves.*) In chorea, the palpitation and difficulty of breathing generally occurs at night, when the patient is perfectly quiet. During the intervals of the attacks, the patients —unless anæmic or disease of the heart exists—can walk fast, and even run, and lie on either side of the chest, without difficulty.

CAUSES OF CHOREA.—Disease of the heart, or its valves, may cause periodical attacks of palpitation, either with or without chorea of some of the muscles of the body

Chorea of the upper part of the body may favor the occurrence of this affection in the heart more than chorea of the lower extremities.

In young men of sedentary habits, sexual abuse and smoking to excess seem to act as excitants. In young females, a disordered state of the menstrual discharge, attended by leucorrhœa and pain in the left side, may excite or predispose to it. From the connection which exists between chorea and rheumatism, it is possible that cases may occur—although no case has yet fallen under my notice—in which rheumatism may excite chorea of the heart.

Chorea of the organ, if of long duration, may produce enlargement and valvular disease.

TREATMENT.—The preparations of iron with phosphate of copper* and choral hydrate being the best remedies for chorea, they seem to be well adapted for this disease when it attacks the heart. When it is complicated with disease of the organ some modification

---

* It is prepared from the phosphate of soda and sulphate of copper. The dose is from 1-12th to 1-6th of a grain.

of this treatment may be necessary. The patient must be cautioned against self-abuse, excessive intimacy, and drinking and smoking. Cold bathing or sponging of the chest and back should be ordered night and morning. If the patient's habits are sedentary, a more active, out-of-door life, will be of great service.

Chlorotic females, when sent into the country and allowed to walk or ride about, instead of being cooped up in close rooms generally recover in a few weeks.

CHAPTER III.

NERVOUS PALPITATION OF THE HEART.

Characters.—Connection with Chorea of the Heart.—Females
more liable than Males.—Causes.—Influence of Age in
producing it.—Symptoms.—Diagnosis.—Treatment.

General Characters.—Nervous palpitation of the heart is
extremely common in the colony. It is very closely allied to chorea
of the heart (*see preceding chapter*), and may end in it. They may
even alternate, the patient at one time suffering from attacks
of chorea, and at another from nervous palpitation. Females are
much more liable than males; this is due to their habits being
more sedentary, and to the connection between the uterine and
nervous systems. If they are single, they are more or less chlorotic; the menstrual discharge is pale and scanty, and more or less
irregular, and attended with pain and discharge of clots of blood
and pieces of membrane, which may form more or less complete
casts of the interior of the womb. The membrane, if the pain is severe,
may become rolled up into little round masses by the action of the
womb, as hard as peas. One female suffering from nervous palpitation passed these masses for three or four months. In water,
they looked like white peas. Leucorrhœa, more or less profuse, is
generally present. Married females are not exempt from this form
of the disease, but they are generally sterile. In females who
have had children it occurs from over suckling, and it is particularly liable to come on in the hot weather, from the menstrual discharge occurring at the same time, or from the existence of

leucorrhœa, whether vaginal or uterine, and disease or prolapsus of the womb. It is more frequently met with in the overworked than in those who are able to take care of themselves. It is very likely to occur at the change of life, and to be accompanied by what females here call "hot flushes, or heats"—see page 2.

Pregnancy, if the patient is feeble and nervous, will be generally found to aggravate the attacks of palpitation.

The general health is sure to be more or less disordered; the lips, gums, and skin, pale; the bowels irregular, usually confined; the urine variable in quantity, either pale and copious, or scanty, and loaded with phosphates. The pulse is feeble, easily quickened by exertion or mental excitement, and the muscles soft. They are generally hysterical.

In males it is generally observed when they are growing fast, and amongst those who are kept closely confined. Self-abuse, excessive intimacy, smoking, drinking, and late hours, have much to do with producing it. Smoking, particularly in hot weather, will often produce and aggravate the attacks. They are generally very nervous, deficient in confidence, the face and gums pale, the flesh soft, and the pupils large. In adults this affection is very often attended with great mental depression, and sometimes, when they have been much addicted to self-abuse, by more or less monomania.

INFLUENCE OF AGE.—This affection is rarely observed before the 23rd year. Of 92 cases which I collected amongst females—

16 were between the 20th and 25th year (inclusive).
18 „ „ 25th and 30th „ „
26 „ „ 30th and 35th „ „
16 „ „ 35th and 40th „ „
2 „ „ 40th and 45th „ „
___
78 cases.

Of the remaining 14 cases, 8 were 20 years of age and under, 2 being 16; the other 6 cases were from 48 to 56 years of age.

From this table it will be seen that this affection is most common when the uterine functions are most active. It is, as I before stated, rarely met with in males, only 18 cases being observed. The youngest ranged from 16 to 20 years, and the oldest from 35 to 40.

It is more commonly observed in private than in public practice.

GENERAL SYMPTOMS.—The symptoms referable to the heart vary but little. The palpitation is most troublesome when some sudden mental shock occurs, such as a knock at the door, or some one entering suddenly; but at times, when perfectly quiet, or when the attention is engaged, or after taking a little wine or some other stimulant, there is but little complaint made of it. The examination of the heart will show that with great excitement and loudness of action there is no alteration of its sounds, beyond occasionally a slight murmur, particularly over the mitral valve, with the first sound. In a lad who has lately been under my care, the murmur over the mitral valve was very strongly marked. It disappears when his health improves, but returns when he loses ground. The cause of the alteration appears to be excessive growth, and masturbation. The murmur when heard is soft and blowing, not hard and rasping, as in organic disease; neither is it continuous, for it may be absent in the morning, but present in the after-part of the day, or after mental excitement or bodily exertion. Some patients, and particularly those suffering from anæmia, often complain of a feeling of oppression or beating in the epigastrum, as if the opening in the diaphragm constricted the ascending vena cava. A blowing sound without harshness, must not be considered as a positive indication of disease of the valves, for it may depend on nervous causes, on alteration in the position of the heart, and even on adhesion of the pericardium to the heart.

The other symptoms are as well marked as those referable to the heart; the face is pale, the muscles flaccid, the conjunctivæ bloodless, the gums the same, and there is often a red line close to the gums, varying in depth and in colour, from pale pink to a deep red, the pulse is either quick and feeble, or very soft; very little exertion fatigues, the urine is pale in the uncomplicated form, and often passed in large quantities, particularly after the attacks.

The symptoms referable to the womb are, either complete suppression of the menstrual discharge, with or without leucorrhœa, or the menstrual discharge is profuse, occurring every 14 or 21 days and clotted or mixed with pieces of membrane. The membrane are more frequently observed in scanty than in profuse menstruation. The pain attending the discharge of these membranes is very often accompanied by severe disturbances of the heart-action. Pain in the left side is very commonly observed in single and in sterile females.

When the palpitation occurs during suckling, leucorrhœa is common, and the menstrual discharge is usually regular, and very often profuse.

In males addicted to self-abuse or excessive intimacy, the same symptoms will be observed as in enervated females, pallidity of the face, nervousness, timidity, and a dislike to come in contact with people, and more or less tendency to melancholy, and the palpitation is easily produced and aggravated by very slight causes acting on the mind.

When this affection occurs in connection with dyspepsia, chronic gastritis, chronic duodenitis or diseases of the liver, womb, or colon, it will be found that the palpitation is always more troublesome when the symptoms referable to these organs are worse. The dyspeptic, and those suffering from chronic gastritis, will complain most during the digestion of a heavy meal, from an aggravation of the pain in the stomach. Those labouring under chronic-duodenitis complain most after the food has left the stomach, and more after farinaceous substances than meat.

In congestion of the liver, particularly in drunkards, with the attacks of palpitation, the heart, they say "rolls over," sometimes the pulse during this will become irregular and feeble, the face pale, and covered with sweat, and they often complain that they feel as if dying. If care is not taken to lie down when this occurs, death may even take place. The same symptoms are often noticed in enlargement of the liver, and in distension of the transverse arch of the colon, either from flatulence, or an accumulation of fæces.

DIAGNOSIS.—It is very difficult to determine whether the palpitation depends on organic disease of the heart or not. It may be assumed that the palpitation does not depend on disease of the organ, if the patient can walk fast or lie on either side, and that under the influence of excitement they can exert themselves as well as when in good health, although not for the same length of time, or without suffering severely afterwards.

It is of the utmost importance that a distinction should be made between nervous palpitation, and palpitation arising from disease. To treat a patient suffering from the former with digitalis veratrium veridium, and other depressants of the heart's action, is a mistake. These drugs will be sure to aggravate the symptoms, and if given too freely even cause a fatal termination. I have seen several cases where the patients' lives have been made miserable and their symptoms aggravated by the injudicious use of these remedies.

It is only by very careful investigation of all the circumstances connected with each case, that a correct diagnosis can be arrived at.

TREATMENT.—The chlorotic and anæmic will be benefited by iron and the phosphate of copper,* change of air, out-of-door occupations, good diet, and a glass of wine two, three, or four times a-day, as the system may bear it. The body should be sponged with tepid or cold water every morning and evening, and well rubbed afterwards with a coarse towel, until a glow is produced. Some patients will not at first bear sponging with cold water. When this is the case, the chill should be taken off the water for some days, or the body well rubbed with a coarse towel or flesh-brush, wetted with cold water. When the patient suffers from womb disease, leucorrhœa, or night emissions, sitting in cold water as high as the navel, for several minutes— two to ten—night and morning, will be found of great service. When the child is very young iron and quinine will cause the mother's milk to disorder its stomach. I have then found the exhibition of an ounce of the decoction of yellow cinchona, with from 3 to 5 grains of carbonate of ammonia, and half a dram of tincture of columbæ, two or three times a day, of great benefit.

If the patient is suckling, the child, if old enough, should be weaned. If the womb is prolapsed, it should be replaced, or if diseased, treated by appropriate remedies.

---

* The solution of the perchloride of iron in doses of from three to five drops, in a little sugar and water, three times a day, is one of the best preparations. The phosphate of copper may be given with the iron, in doses of from 1-12th to 1-6th of a grain, made into a pill, with extract of gentian.

Palpitation depending on chronic gastritis is best relieved by careful dieting—cocoa, chocolate, or milk and water, being taken instead of tea and coffee, and not in large quantities at a time, and dry toast, instead of butter and meat; for dinner, beef tea and fish, or a lean chop, with one kind of vegetable, with dry bread and a little sherry or weak brandy and water. They should rest from half to an hour after dinner, for nothing tends so much to disorder the digestion as active exertion on a full stomach, particularly if the food is swallowed in a half-masticated state. They should be very careful not to eat too much.

When the palpitation depends on disorder of the duodenum, or liver, or distension of the colon, aperients, particularly the watery extract of aloes, in doses of from 1 to 2 grains, with 1 grain of Castile soap, may be taken every or every second night, to insure an evacuation from the bowels every day. Great benefit is generally experienced from the action of the aloes on the rectum, if not given too freely to irritate it. If there is any tendency to jaundice or bilious vomiting, the addition of a sixth or a fourth of a grain of podophylline to each dose of the aloes, will be of great service. The diet must be carefully regulated, and the body sponged night and morning, either with tepid or cold water.

CHAPTER IV.

INFLAMMATION OF THE PERICARDIUM.—PERICARDITIS.

GENERAL OBSERVATIONS.—RELATIVE FREQUENCY OF THE VARIOUS FORMS.—INFLUENCE OF THE SEASONS IN CAUSING THIS AND OTHER DISEASES OF THE HEART.—FREQUENCY.—GENERAL SYMPTOMS.— TERMINATIONS.—CASES.—AFTER DEATH APPEARANCES.—COMPLICATIONS.—TREATMENT.

GENERAL OBSERVATIONS.—Inflammation of the pericardium sometimes occurs here without being complicated with any other disease. It is, however, but very rarely observed, from the heart and its covering being well protected under the ribs. The most frequent causes are—1st, rheumatism of the upper extremities, particularly of the shoulders and neck, and of the diaphragm and intercostal muscles; and, 2ndly, inflammation of the left lung and left pleura.

The pericardium, and with it the valves, may become affected at the very commencement if the rheumatic attack is very severe, and the patient at the same time exposed to cold and wet, but generally the rheumatism exists for several days, and is more or less erratic and liable to be aggravated from exposure to cold or wet, the shoulders, neck or chest being affected, or it is fixed in one of the shoulders, generally the left, and creeps in a distinct manner to the fascia of the neck, and thence down the prolongations which it sends into the chest and along the sternum, until it reaches the pericardium and the valves of the heart. In the same way in rheumatism of the diaphragm and intercostal muscles the pericardium may become implicated.

In inflammation of the left lung, or the left pleura, the pericardium rarely escapes being more or less implicated, but from the valves not becoming so often affected as in rheumatism it generally escapes notice, unless the effusion into the pericardium is considerable. Even in these cases, unless the history is very carefully traced, it may not be discovered.

RELATIVE FREQUENCY OF THE DIFFERENT FORMS OF PERICARDITIS.—Of the 45 cases collected, in

5 it was primary and uncomplicated.

11 complicated with pleuritis of the left side, with inflammation of the left lung.

29 with rheumatism, either acute or chronic.

INFLUENCE OF THE SEASONS IN CAUSING PERICARDITIS AND DISEASES OF THE HEART.—If the weather is wet and cold, and particularly if it is changeable, passing from extreme heat to extreme cold, then inflammation of the lungs and rheumatism, with disease of the heart, is sure to be met with. The weather in the winter and spring of 1870 was very cold, wet and variable. A larger number of cases than usual sought relief at the Institution.

Extreme heat and extreme cold often act very injuriously on heart diseases. Old cases may sink or become worse during hot weather from exhaustion or congestion of the lungs; and cold weather may bring on another attack of rheumatism, or excite bronchitis or congestion of the lungs.

The following table, compiled from the Registrar-General's returns, will show the number of deaths registered in Melbourne, for twelve months, from pericarditis, disease of the heart, aneurism and rheumatism ; and the mean temperature during the first and fourth week :—

Months.	Peri-carditis.	Heart Disease.	Aneurism.	Rheuma-tism.	All Causes.	MEAN TEMP. OF	
						1st Week of Month.	4th Week of Month.
March	1	3	...	...	337	77°	59°
April	...	7	4	...	346	64°	44°
May	...	3	...	2	337	54°	49°
June	3	3	3	...	337	47°	50°
July	3	6	1	2	265	47°	45°
August	1	5	...	...	265	45°	51°
September	1	10	2	...	234	51°	56°
October......	...	5	2	1	226	57°	59°
November...	...	8	...	1	186	56°	58°
December...	...	5	5	1	322	61°	64°
January ...	...	5	3	...	401	65°	68°
February ...	1	12	...	...	335	65°	67°

On referring to the third annual report of the Registrar-General of England I find the following table of the mortality from disease of the heart and rheumatism :—

	Winter Quarter.	Spring Quarter.	Summer Quarter.	Autumn Quarter.
Disease of heart	739 ...	556 ...	571 ...	698
Rheumatism ...	124 ...	113 ...	99 ...	117
Total ...	863	669	670	815

From this table it will be seen that the number of deaths is greater in the winter and autumn quarters than in the spring and summer.

In Melbourne the deaths from heart diseases in the winter and summer quarters are nearly equal—

AUTUMN QUARTER.

	Heart Diseases.	Rheumatism.
March April May } were ...	14 ...	2

WINTER QUARTER.

Heart Disease.   Rheumatism.

June
July   } were   ...   21   ...   2
August

SPRING QUARTER.

September
October   }   11   ...   0
November

SUMMER QUARTER.

December
January   }   23   ...   0
February

FREQUENCY OF PERICARDITIS.—This has been alluded to in the first chapter. The number of admissions into the hospitals here, as shown in the statistics, are fewer than in England. This arises not from the disease being less frequent, but from its being of a less active character, and occurring later in the rheumatic attack than in the old country. From effusion taking place very early in the attack, the rough sound caused by the congested membranes rubbing against each other may disappear in the course of a few hours, and then if there is no complaint made of pain in the chest, the existence of pericarditis is unsuspected.

From 1857 to 1869 inclusive, there were only 25 cases of pericarditis, according to the reports, admitted into the Melbourne Hospital. This refers to cases presenting symptoms of pericarditis when admitted, and entered as such in the books. During nine months I noted a larger number of cases of rheumatic pericarditis in the wards, and scarcely any cases of acute rheumatism of the shoulders occurred without the heart being more or less affected.

The deaths stated to have occurred in the Hospital returns from rheumatism were generally from diseases of the heart.

In 1857, of 135 cases treated, 11 died ; but from this time to 1865, although the number ranged from 113 to 256, the deaths were not numerous; but from 1865, when out of 302 cases treated 6 died—7 out of 185 in 1867—8 out of 233 in 1868, and 7 out of 252 in 1869.

GENERAL SYMPTOMS.—1st, of the uncomplicated form.—From exposure to cold and wet, the patient is seized with chills followed by hot flushes, thirst, white tongue, with more or less acute pain in the side. With the pain, which may be so slight as to escape attention, and supposed to be inflammation of the pleura, there is the peculiar friction sound caused by the congested lining membrane of the pericardium rubbing against that portion of the membrane covering the heart. The pain when only the pericardium is affected is confined to a space of the size of the palm of the hand, and each motion of the heart aggravates the pain by bringing the inflamed membranes in contact. The pain differs from that observed when pleuritis exists. It is not aggravated by breathing. From serous effusion taking place rapidly, the friction sound may disappear in from 12 to 18 or 24 hours, and mask the symptoms of pericarditis. By the following simple plan it is generally easy to detect the presence of fluid in the pericardium, and determine the cause of the increased dulness on percussion according to the amount effused, and the feebleness of the heart sounds. If the patient is either placed with the body bent forward in the sitting position, or the left side brought over the side of the bed, the heart, from being heavier than the fluid effused, falls against the pericardium; the friction and heart sounds can be then heard as plainly as if no fluid existed. The value of this simple plan was strongly shown in two cases—one of cancer of the pericardium, the other in which an hydatid cyst existed in the lower lobe of the left lung, The pain, and the increased area of dulness on percussion, with some obscure friction sound, led to the supposition that there was pericardial inflammation. Changing the position of the body did not, however, bring the heart nearer the ribs. The post-mortem examination of the cancer case showed that the pericardium was adherent to the lower lobe of the left lung; they formed a cancerous mass nearly as large as two fists. This mass had pushed the heart back. The patient was unable to lie down, and the pain was most severe at night. In the case of the hydatids, after spitting up a large quantity of blood, a large number of cysts were expectorated, the heart resumed its natural position, and the patient recovered. From the obscurity of the symptoms a large quantity of serum is sometimes found in the pericardium after death, its existence being unsuspected during life. In a case of this kind, in which severe pain in the side followed bathing in a waterhole on a very hot day, Dr. Hudson found several pints of serum. The following case bears some resemblance to it:—

The patient, a clerk aged 23, had got wet, and remained in his wet clothes for several hours. He was taken with cold chills followed by heats and pain in the left breast. He remained in this state for more than a fortnight, when he was seen by a medical man. He so far recovered under the treatment employed as to be able at the end of six weeks to walk about. He was thought to be consumptive, as there was dulness on percussion on the lower part of the left lung, with cough, expectoration, and night sweats. When seen, he presented the following state :— Pulse 120; unable to lie on the right side or back; urine scanty and high coloured; irregular sweats preceded by chills; cough with mucous expectoration; emaciation marked, and tongue slightly coated. Mucous râles existed in the front of the upper third of the left lung, with slight dulness on percussion, but pos-

terially the whole length of the lung was dull—it was most marked
below the spine of the scapula, and was attended with slight nasal
voice sound. The dulness on percussion in the front of the chest
in the region of the heart was strongly marked over a space nearly
the size of the two hands; below it blended with the left lobe of the
liver, which was pressed below the false ribs. He felt, he said, as
if a heavy stone was on that side of the chest; and if he laid with
his head low, it seemed to shift up, and threaten to suffocate
him. He could only lie on the left side with the shoulders raised,
and the chest and body drawn together. The heart's sounds were
inaudible, but by placing him on the left side, with the chest in
the lowest position, the heart's sounds were brought into notice.
By repeatedly blistering the chest, and giving him small doses of
calomel and opium, and iodide of potash, the fluid was absorbed in
the course of four or five weeks.

*When complicated with pleuritis or inflammation of the left
lung.*—There is, as I have observed, a very great tendency on
the part of the pericardium to become implicated in pleuritis,
and in inflammation of the lower part of the left lung. The
symptoms indicating the pericardial implication are often so obscure,
the pain being masked from being less severe than that which
exists in the pleuritis, and by the more extended friction sound,
that unless the state of the heart is examined very early, before
effusion takes place, and while the breath is held, its existence
can scarcely be detected. If the effusion is considerable, then the
increased area of dulness in the region of the heart, the feebleness
of the heart's sounds, rendered more prominent by causing the
patient to lean forward, and the peculiar position which the patient
chooses, which will be generally more marked when the effusion

takes place rapidly than when it takes place slowly—namely—
lying on the left side, with the chest bent forward to allow the
heart to take from its greater weight the lowest position, and prevent the root of the lung being pressed upon.   If there is any
doubt as to the nature of the effusion, although no better
proof can be needed than with the increased extent of the dulness on percussion, and the heart's sound being made clearer or
feebler by alternately bending the chest forward and back-
ward; placing the shoulders lower than the abdomen will excite
symptoms of suffocation, from the heart being pressed against the
root of the lung.

In most of the cases of this complication which have fallen under
my observation the patients were young.   They either resided in
damp and draughty houses, or were exposed to sudden alternations
of temperature.   It was seldom complicated with rheumatism,
unless they were beyond the eighth or tenth year.   The usual history of these cases, when met with amongst the poor and badly
housed, during the winter months, no matter whether the weather
is wet or dry, but more especially if the days are hot and the nights
cold, is as follows:—The child is seized with inflammation of the left
lung.   It gets better, and from cold gets a relapse; the other lung
may become affected ; and the child, weakened and emaciated by
the previous attack, dies, or the lung symptoms improve so much
that the state of pallor and emaciation which continues can scarcely
be accounted for.   The skin is always cold, the muscles flaccid,
the sub-lingual temperature is from three to four or even more
degrees lower than in a child of the same age in good health.   The
thermometer rises very slowly both in the palm of the hand and
under the tongue.   There is generally an access of fever at night.

It may linger on in this state until the hot weather sets in, when symptoms of softening of the stomach is liable to occur. Then the milk is vomited in a curdled state, the motions are of the same nature, and more or less mixed with green and yellow bile, or thin and watery, mixed with little specs of green bile like minute pieces of parsley, and staining the napkins like cabbage-water.

Children who suffer from lung disease in the winter are very liable to be attacked with symptoms of softening of the stomach in the hot weather, and if their strength has been very much reduced it is very apt to prove fatal.

The following are the best of the cases collected :—

Inflammation of the left lung, with effusion into the pericardium.
—Sudden congestion of the right lung.—Death.

A child aged eighteen months was brought to the Institution suffering for three or four days from cough and difficulty of breathing. The face was pallid; lips bluish; pulse 150, feeble; skin cold; and the emaciation extreme. There was scarcely any respiratory sound to be heard in the left lung; but in the right one there was extensive crepitation. It died the same night. There was nothing in the previous history of the case to lead to the suspicion that there was any pericardial disease. About three months before it had had an attack of inflammation of the left lung, for which it had been an out-patient of the hospital, but the cough, which had never been very severe, remained. It gradually lost flesh, and became paler. The house was cold and damp. Another child, five years of age, had died from consumption.

The pericardium contained nearly half-a-pint of fluid, which was

D

slightly opaque from the presence of flakes of purulent matter. The membrane covering the heart and the pericardium was of a bluish colour. The left lung was indurated, and slightly adherent at its lower part. The right lung was congested; this was recent, and the cause of death. The heart was very small.

Here, in disease of one of the lungs, if the patient is much reduced, the occurrence of congestion of the other is very apt to be followed by a fatal result.

In another case, a child two years of age, death ensued from softening of the stomach; the left lung became inflamed, from exposure to cold. Several relapses occurred, and sometimes the right, sometimes the left lung was affected. The disease yielded to treatment, but the child remained pale and emaciated. It continued in this state for two months. The occurrence of two very hot days in October excited symptoms of softening of the stomach, and it died from exhaustion on the seventh day. The mucous membrane of the stomach was softened. The lungs were healthy. The pericardium contained about six ounces of slightly opaque fluid. In the following case the symptoms were very acute, death ensuing in ten days. The patient, seven months old, was exposed in the month of September, after a very hot day, to the cold night air. In the night it had an attack of croup, with difficulty in breathing, attended with considerable wheezing. It was supposed to be better the next day, from the exhibition of Ipecacuhana wine, and linseed and mustard poultices to the chest and back. On the seventh day the child was brought to the Institution. Its face was pallid; lips congested; pulse 120; skin cold; bowels rather open; urine high coloured. Fine crepitation existed in both lungs, but in the left it was rather coarser than in the right one. The heart was acting with unusual force. The child laid on the left side, with the chest bent. When it was lifted up or straightened, it began to cry and continued to do so until it was replaced on the left side. The tumultuous action of the heart, and the position it was easiest in, pointed to some cardiac disturbance, but no alteration could be discovered. During the next three days there was but little alteration in the symptoms. On the tenth day convulsions set in, and it died in the course of a few hours.

The head was not allowed to be examined. Both lungs were very much congested, particularly the lower part of the left one. There were three ounces of serum in the left pleural cavity. The pericardium and the surface of the heart were deeply injected. The pericardium contained four ounces of serum. The lining membrane of the heart and the valves were injected, but there was no deposit of lymph.

In the next case, the symptoms of pericarditis were quite as obscure as in the other cases. The heart became dislocated, and from bending of the aorta, symptoms of valvular disease were excited, and then dropsy.

A male aged twelve years, residing in an old wooden house in Romeo-lane, was seen October the 5th, suffering for one month from pain in the left side of the chest, with cough and expectoration of thick mucus. At the commencement, the phlegm was tinged with blood. The skin was hot and dry; the pulse 120, rather full; tongue white; urine high coloured; face pallid, emaciation extreme, and increasing. He had no distinct attacks of night sweats although he was very feverish, and at times broke out into sweats, which were rather saltish to the taste. There was increased dulness on percussion in the whole of the left lung, absence of respiratory sound; but in the course of the bronchial tubes, particularly the large ones, coarse râles both in breathing and coughing existed. The heart presented no change, although at the commencement there was pain over it, and he continued to complain, although less severely. He preferred lying on the left side, with the chest bent forward. He steadily improved under treatment, the phlegm diminished, and the respiratory sound re-appeared in the lung, and beyond being a little weak and unable to walk fast, from some trouble about the heart, he seemed to be recovering.

Nov. 16th—Looks better and has gained flesh and strength. The last few days he has complained of more pain, and beating of the heart. On examination, the base of the heart was discovered just above the left nipple. There was a murmur over the mitral and aortic valves. From this time until his death, the base of the heart gradually rose as high as the lower border of the third rib, and the murmurs in the mitral and aortic valves increased in loudness; the liver extended below the ribs. Dropsy set in slowly. Diuretics and tapping were had recourse to, without checking it. The last time he was tapped only one pint of fluid was obtained. He sank from exhaustion and diarrhœa. The heart was very much enlarged. Its base was directed to the left, and was opposite the second bone of the sternum; its apex was on a level with the fourth rib. There was no alteration either in the mitral or aortic valves. There was no pericardium, save at one small space, near the origin of the aorta, of the size of a five-shilling piece. The lungs were healthy, and the liver small. The abdomen contained several pints of serum, which was bound in by adhesions between the peritoneum and the small intestines, into five or six sacs, containing from half to a pint of fluid. When last tapped the trocar, a very fine one, had penetrated one of these cysts.

3. SYMPTOMS OF RHEUMATIC PERICARDITIS.—The pericardium, as before observed, may become implicated at the onset of the rheumatic symptoms, if they are severe and general, but more frequently the pains are erratic, passing from the hand to the shoulder, and even to the diaphragm, or from the ankle to the knee or the hip; every additional chill adding to the severity of the symptoms. When the lower limbs are affected there seems to be very little liability on the part of the heart and pericardium to become implicated ; but when the upper, and especially the shoulders and neck, then it will be very strong. The same tendency will be observed when the diaphragm is affected.

In England rheumatism generally attacks the large joints, but here, although it often does the same, there is seldom very much swelling, unless the patient is very muscular, in the prime of life, and has been very much exposed to cold and wet. The type of the disease is as a rule much less imflammatory than in England. It is not fixed to the fibrous tissues of the joints, but attacks with nearly equal frequency those muscles which contain a large quantity of fibrous tissue, and the fibrous tissues or fascia.

The following case will illustrate how rheumatism often extends :—The patient, a professional man, had suffered from pain of a rheumatic nature in the right index finger and wrist; it alternated with pain in the intercostal muscles and the diaphragm. When it was severe it caused some obstruction to the descent of the first two or three morsels of food through the diaphragm. From being too thinly clad while riding in a cab the pain in the lower part of the chest became much worse, he had troublesome hiccough, the skin became slightly jaundiced, urine dark-coloured and loaded with lithiates, skin dry and harsh, with frequent hot flushes of a passing character, ending in cold sweats. The sweat was sometimes acid to the taste, sometimes saltish, and sometimes insipid. These differences in the taste seemed to depend on the intensity of the fever and the quantity of perspiration; when the former was high and the latter scanty, then it was observed to be more or less acid; when more profuse, it was alkaline; and when still more profuse, tasteless. The pain in the diaphragm was never fixed; it alternated with pain in the *right* wrist and left shoulder. From fresh exposure to cold the pain settled in the left deltoid muscle and under the shoulder blade; in the course of two days it extended from under the shoulder blade along the clavicle to the fascia of the neck and the under surface of the sternum. At the same time there was pain in the chest, behind the sternum, which each beat of the aorta and each inspiration aggravated. About the third day there was pain in the left side of the chest, occupying a defined space somewhat less in size than the palm of the hand. The impulse of the heart aggravated this pain, and pressure with the finger between the ribs excited pain. With the pain in the pericardium a slight murmur was heard in the mitral and aortic valves, and some roughness in the pericardium.

It very seldom happens that the extension of the pain can be so accurately traced, as in this case, from the shoulders and neck down to the pericardium and the valves of the heart. I have so often seen disease of the heart occur in connection with rheumatism of the shoulder, and particularly the left one, that I have always looked upon its existence as a very strong indication of a tendency to disease of the former. The same observation applies to rheumatism of the diaphragm, which is, however, from being so well protected, rarely affected. A short time ago I had an opportunity, through the kindness of my late dear friend, Dr. Stewart, of seeing a lad, aged 16, a shoemaker, in whom the pain in the diaphragm was marked, there being no rheumatism of the limbs. He had had two years before rheumatism, with palpitation of the heart, and was laid up for a month. The palpitation got better after he got about, and he has since been able to work and run without suffering from it. Sixteen days ago he ran some distance, and sat for some time on a cart, exposed to cold and wet. He was soon afterwards seized with chills and pain across the lower part of the chest, followed by palpitation of the heart. He has now no rheumatism, but the pain in the diaphragm continues; the dulness over the region of the heart is somewhat increased in extent, but no friction sound or alteration in the sounds can be heard by altering the position of the chest. There is a loud, but not harsh murmur heard over the mitral valve, and the murmur can be heard, but not so loud along the lower border of the scapula. There is also a murmur at the base of the heart, extending up along the aorta. The pulse is 106, without much power; tongue white, and skin harsh. He had considerable amount of fever at first, but it had diminished under the treatment which had been employed. He subsequently entered the hospital.

Pericarditis does not seem to be, if the patient is protected from cold and properly treated, liable to prove fatal, unless the effusion is considerable, or disease of the mitral and aortic valves occur— a frequent complication; then from the circulation of the liver becoming affected, dropsy will set in. From the skin being

much more active here than in England, the kidneys are not
nearly so often implicated ; the urine, therefore, seldom becomes
albuminous. When pericarditis occurs after scarlet fever, a not uncommon result, the urine will generally be found to be albuminous.
The same may occasionally happen after measles. The usual history of these cases are, that while the skin is desquamating, from
exposure to cold, symptoms of congestion of the lungs, kidneys,
peritoneum, and pericardium set in. The urine is either very
scanty and albuminous, or altogether suppressed. When this
occurs they may die from uræmic poisoning. There will be more or
less serous effusion into the ventricles and the membranes of the
brain ; the lungs, liver, and spleen congested ; the pleural and
abdominal cavities and the pericardium will contain fluid. These
were the *post mortem* changes found in a child three years of age,
who from cold, while the skin was desquamatory after scarlet
fever, died comatose in seventy-two hours.

In the following case the symptoms occurred after measles :—
An infant, aged eighteen months, had had measles, which was
then very prevalent. It seemed to do well up to the tenth day,
when, from exposure to cold after a very hot day, it became
feverish and restless, and refused to lie on the left side. The
pulse was 136, very feeble; the urine was scanty, a little collected showed slight traces of albumen. There was distinct crepitation in both lungs; the pericardial region was very prominent,
and the child cried when it was touched, or the body straightened ; the heart sounds were feeble, but became louder when
the left side was the lowest point. It preferred the left side, and
laid with the chest bent forward, and the abdomen and legs
drawn up. Blistering, calomel, and Dover's powder, with diuretics,

were ordered, without any benefit. The effusion became so marked that the entrance of air into the left lung was interfered with. It became convulsed and semi-comatose on the fifteenth day and died. The pericardium contained four ounces of reddish serous fluid. The surface of the heart and the pericardium were injected. Both the pleural cavities and the abdomen contained serum ; the kidneys and lungs were congested. When life is prolonged, lymph and purulent matter may be expected to be found, as in the following case :—A delicate cachetic girl entered the hospital on the 12th of May. She had first measles, and then scarlet fever. From exposure to cold two or three days before admission, while the skin was desquamating, she began to complain.

On the 15th, she presented the following state : No urine passed since admission; she is extremely fretful; cries when touched; the cries are redoubled when the region of the kidneys is pressed, particularly the left one; tongue red, with several small sores on it; pulse 102, rather full; left elbow swollen; both legs drawn up, and, when they are straightened, the cries are increased in intensity; the right knee appears to be a little swollen; respiration a little harsh, with slight mucous wheeze. She had a warm bath and diuretic mixture.

16th.—She was so very restless on the previous night, that oneeighth of a grain of morphine was given—this caused her to sleep. The pupils were contracted; pulse 102, not quite so full; no urine has been passed ; bowels confined ; cough rather troublesome ; mucous râle all over the anterior part of the chest, the posterior part could not be examined ; during the examination she vomited some greenish fluid, about half a teacupful; tenderness existed equally over both kidneys, for when the loins were pressed, she uttered louder cries ; heart's action laboured ; a slight friction sound observed. • Compound jalap powder was ordered every three hours until the bowels acted, with super-tartrate of potash for drink.

18th. Skin cool ; pulse 108, feeble; cough frequent ; tongue coated with a little brown fur; passed a large quantity of urine in the night, and vomited some more greenish fluid; bowels acted after the third powder, assisted by an enema; elbow better, and she does not cry when her loins are pressed, but she is very fretful; takes only a little wine and water; urine passed under her; there are mucous râles at the base of the chest, both anteriorly and laterally, but superiorly there are sibilant râles. She died in the night.

Mr. James kindly furnished the following notes of the *post mortem* examination. The pericardium contained two ounces of serum ; its lining membrane was congested. There were deposits of lymph between the stomach and the spleen, and between the rectum and the uterus ; the superficial layers in the former situation had run into suppuration ; the lungs were highly congested ; the kidneys the same—in the left one there was a cicatrix of some standing, as if there had been an abscess; there was a similar cicatrix in the spleen, under the lymph ; the spleen was gorged with dark blood ; the liver was congested; the duodenum and upper part of the ileum were coated with bile ; the large intestines and the rectum were covered with small ulcers; the mesenteric glands were enlarged; there was very little blood in the vessels—it was pale and watery.

AFTER DEATH APPEARANCES.—The changes found after death differ somewhat from those found in England, from the disease being less acute. Extensive deposits of fibrine are sometimes found in strong robust men and females who have been exposed to cold and wet; but adhesions which may consist of fine cellular bands of variable length and fineness uniting the pericardium and the heart more or less intimately, are more common. Effusion with or without flakes of lymph, and congestion of the covering of the heart and pericardium, but rather of a venous nature, are also met with. From the disease not being of a very active nature, it rarely proves fatal, and therefore few opportunities occur of observing the *post mortem* appearances. From its insidious nature, it is often unsuspected during life, death from some other disease bringing to light the fact that there had been pericarditis at some time or another.

The following cases are good illustrations of rheumatism and rheumatic pericarditis as observed in the colony —

Acute Rheumatism affecting the Ankles, then the Knees, and lastly the Wrists and Shoulders, and extending along the Fascia into the Chest, causing Inflammation of the Pericardium.

A female, aged 19, a domestic; eight months in the colony. She was strong and healthy-looking, and had always enjoyed good health. Family healthy. Six days before she came under notice she was seized with pains in the ankles, from wetting her legs and feet. On the second day her knees were affected, and by the fifth her wrists, and then the shoulders. There was not much swelling of the parts; heat of skin not much increased, but the skin felt greasy; pupils dilated; headache; pulse 96, without much power; urine natural. The existence of the rheumatism in the shoulder led to the state of the heart being carefully examined, but no alteration could be discovered; its sounds were somewhat muffled; dulness, 4½ fingers broad from above, downwards, and 4 fingers transversely. Respiration was natural, but she complained, on drawing a full breath, of pain in the upper part of the chest, referable to the arch of the aorta. There was general tenderness of the fascia of the shoulders and neck. Ordered colchicum with iodide of potassium and henbane, every four hours, with morphia and extract of colchicum at night.

7th day, 8 a.m.—Face anxious; eyebrows knitted; complaining of pain in the left breast, passing up and back to shoulder; palpitation of the heart, and difficulty of breathing; pulse 130, hard and regular; heat of skin slightly increased; cheeks flushed and covered with perspiration. She had awoke at 2 a.m., with great difficulty of breathing and palpitation. A mustard poultice relieved the difficulty of breathing somewhat. She is unable to take a deep breath from a sense of constriction across the chest; heart's impulse short and jerking; no alteration in the valvular sounds, but there is a double friction sound heard below the nipple; there is tenderness on pressing between the ribs here; pains in the joints the same; urine natural. Twelve leeches were applied, followed by a hot bran poultice to the whole of the chest; two grains of calomel, with half a grain of morphine ordered, and repeated in six hours; colchicum mixture continued.

8th.—Marked relief to the breathing and palpitation followed the leeches and bran poultice; night tranquil; the friction sound had diminished very much in intensity; pulse had sunk to 110 and was softer; heart's dulness, which on the 7th was diminished in extent, was now the same as on the 6th day; pains less ; urine contained some lithiates ; gums slightly spongy; feels weak. Beef tea ordered.

12th.—Friction sound had disappeared; pulse 90, feeble, but regular ; pains in the limbs much better.

The progress of this case was satisfactory. She had no relapse, a very common occurrence here. This was due to placing her in flannel at once, and to her being well lodged and cared for As a rule, private patients suffering from rheumatism and inflammation of the pericardium, or of the lining membrane of the heart, do much better, and are less liable to relapses than hospital patients. This, of course, only applies to those who are well lodged and well nursed.

This case illustrates the value of energetic treatment at the commencement of the symptoms of pericardial implication.

Acute Rheumatism, affecting first the Lower then the Upper Extremities.—Congestion of the Heart, with Pleuritis, the former increasing.—Later Pericardial Friction and Valvular Alteration.

A male, aged 24, a labourer, one year in the colony. He is healthy-looking, and has always enjoyed good health—habits temperate. Six days before seen, from exposure to wet and re-maining all day in his wet clothes, he was seized with severe pains in the lower limbs and joints, and in the back. After going to bed on the second day, the upper limbs became affected. There was not much swelling of the joints, but they were very tender; the fascia of the limbs were also tender ; skin rather hot and dry ; pulse slow; no sleep since the attack, from the severity of the pain, which was aggravated by warmth ; urine free from sediment;

heart sounds muffled, but no alteration in them; there was friction sound heard below the nipple, and to the left, but it disappears on holding his breath, and below the angle of the scapula there was slight nasal voice resonance. Colchicum with iodide of potassium and henbane every six hours, Dover's powder at bedtime.

In the evening he was seized with difficulty of breathing; anxiety and pain about the region of the heart, with palpitation, flushing of the face, and knitting of the eyebrows. The attack had set in gradually, commencing at seven o'clock, with a sense of suffocation, followed by palpitation, and when seen at ten o'clock had become very severe. The most careful examination failed to detect any friction sound in the pericardium; the heart's sounds were muffled and its action abrupt. Fifteen leeches were applied and the bleeding encouraged by a large, hot bread poultice, and two grains of calomel with one of opium given. At twelve p.m. the leeches had bled freely, the anxiety and palpitation had diminished; still there was no friction sound.

7th.—The heart's sounds could be more clearly distinguished; a friction sound could be detected, and the pain was rather more marked; there was also a slight murmur heard over the mitral and aortic valves; pulse 120, not very strong. Ten leeches were applied, and the calomel and opium repeated.

8th.—The friction sound and murmur less; a blister was applied, and kept open with savine ointment.

9th.—Gums tender; complains of feeling weak; pulse 100, feeble but regular; urine containing a little red sediment. The friction sound had disappeared, but there was still a murmur over the mitral and aortic valves; over the latter it was floppish with the first sound, but slightly crepitating with the second; over the former, both with the first and second sound, it was crepitating. The rheumatic pains had greatly diminished in severity. Beef tea and wine were ordered.

The progress of the case was very satisfactory for the next seven days; the rheumatic pains disappeared and the murmurs diminished to a slight roughness; then, from exposure, he had a slight return of the rheumatism, and with it the murmurs returned over the valves, but there was no return of the pericardial symptoms. The valvular murmurs never quite disappeared, although he got quite well.

In a case which I had an opportunity of watching in the Hospital, the patient, a male, about 20 years of age, three years in the colony, was complaining on the fifth day of general rheumatic fever, of pain below the left nipple on drawing a deep breath; the heart's sounds were muffled, and the pulse slow. In the night he was seized with difficulty of breathing, with anxiety and pain in the region of the heart. The application of leeches relieved these symptoms. The next day, the sixth of the rheumatic attack, the pain was less, but the heart's sounds were still muffled; more leeches were applied, and Dr. Murray, under whose charge the case was, ordered him minute doses of calomel and opium. On the seventh day the heart's sounds were less muffled; friction sound could be distinctly heard, but it was not strongly marked; there was also a feeble murmur heard over the mitral and aortic valves, loudest over the latter. A blister was applied. On the eighth day the friction sound had disappeared; his gums were slightly spongy. He recovered after several returns of the rheumatic pains. The valvular sounds remained floppish. He had had several attacks of rheumatism both in England and in this colony, but this was the first time that he had suffered from pain in the chest and difficulty of breathing.

Rheumatism, Pericarditis, Pleuritis, and Congestion of the Left Lung.

Servant, aged 20, well nourished, 17 months in the colony, seen September 25th, suffering from general rheumatism, but only the left wrist and foot were a little swollen. The pain she said shot through her chest, arms, and legs. It was very severe in the region of the heart, the epigastrium, and the whole of the left side. When desired to hold her breath, the pain was then most marked in the region of the heart; but when desired to breathe deeply the pain shot up to the left shoulder. Pulse 130, feeble, face flushed, skin covered with acid perspiration. She could not lie down. There was marked dulness in the pericardial region; the heart's sounds were inaudible until the body was bent forward. Crepitation and nasal voice sound existed in the left lung as high as the scapula. She had been ill for twelve days, and for seven had been confined to bed. Blister; calomel and opium.

26th.—But little alteration; pulse had increased to 135, lips bluish, crepitation in the upper part of the left lung. Twelve leeches were applied to the chest, followed by a large linseed meal poultice. Iodide of potash, with colchicum and bicarbonate of potash, ordered.

27th.—The leeches bled freely and gave her great relief; pulse 130. The blister to be dressed with mercurial ointment.

October 4th.—Her mouth had been sore for several days. On its occurrence the symptoms gradually improved. It was necessary to give her beef tea and brandy from the commencement. The value of the abstraction of blood was strongly shown in this case; the pulse lost its quickness and became stronger. Prior to the bleeding the opium given with the calomel seemed to aggravate the symptoms. Patients, as a rule, bear large doses of opium well in the colony; 1 grain to $\frac{1}{4}$th of a grain of calomel seems to act better than a 4th of a grain to 1 grain of calomel, and there is less risk of severe salivation.

COMPLICATIONS.—The most constant complication of the rheumatic form is disease of the mitral or aortic valves, sometimes of both. See *Disease of the Valves.* In the chronic form, from obstruction to the circulation, either from adhesions between the pericardium and heart, or from dislocation of the heart, or disease of the mitral valve, dropsy is of common occurrence. The urine, as before observed, is rarely albuminous, from the skin being more active there than in the old country. When observed it is chiefly amongst those who have come from New Zealand, or the colder parts of the colony, or in drunkards who have been much exposed to the weather, and in pericarditis occurring with rheumatism after scarlet fever, and occasionally after measles, from the action of the skin being checked.* In one case there was gonorrhœal rheumatism with the pericardial inflammation. This unusual complication occurred in a young man, aged 19. Attacks of hæmorrhage from the bowels, stomach, and lungs sometimes occur. Some cases were complicated with chronic diarrhœa. Since the extension

* Albuminuria is very common here after scarlet fever, and is occasionally observed after measles. The rarity of albuminaria would indicate that cases of the chronic or incipient form would be benefited by a residence in the colony.

E

of the Yan Yean, from the water drank being purer, this has not been so common.

TREATMENT.—Very active treatment—particularly bleeding, either from the arm or by cupping or leeching the chest, and the free use of calomel—is rarely needed in colonial practice. The practitioner must, however, be guided by the state of the pulse, whether full and hard, or feeble, and the aspect of the patient, whether muscular and strong, or pale and flabby. A robust patient, with a quick, hard pulse, will be benefited by bleeding from the arm, or from 12 to 18 leeches to the chest, and the exhibition of one or two grains of calomel, with from a quarter to half a grain of opium every three or four hours, until the mouth is rendered sore. Patients here do not bear mercury well. They become salivated very quickly, and often unexpectedly; it reduces their strength so much that quinine, beef tea, and stimulants become necessary very early. In most cases a large blister over the heart, with small doses of calomel and opium, iodide of potash and colchicum, with potash or soda, when the perspiration is acid, quickly removes the symptoms. The patient should be placed in a room which should be kept day and night at a temperature of 60 deg. Fahrenheit, and well protected from draughts. The body should be rapidly sponged, either with a warm alkaline or acid solution* every six or eight hours, according to the relief given, and afterwards wrapped up to encourage moderate perspiration. The patient's strength should be kept up by at least the strength of two pounds of beef, made into beef tea, in the

---

* A piece of washing soda, the size of a walnut, in one or two quarts of warm water, make an excellent alkaline wash; and a quarter of a pint of pale vinegar to one quart of warm water, an acid one.

24 hours, and as much wine and spirits as the previous habits of the patient and the strength of the pulse demand. The disease is attended with so much debility, that if the strength is not carefully kept up the convalescence is generally very long, and there is, from very slight cold, from any sudden change of temperature, great danger of a relapse.

CHAPTER V.

DISEASE OF THE VALVES OF THE HEART.—INFLAMMATION OF THE LINING MEMBRANE OF THE HEART.—(*ENDOCARDITIS*).

GENERAL OBSERVATIONS.— FREQUENCY OF IMPLICATION OF THE DIFFERENT VALVES, WITH AND WITHOUT HYPERTROPHY OF THE HEART.—DISEASE OF THE AORTIC VALVES—SYMPTOMS—CASES.— DISEASE OF THE MITRAL VALVES—SYMPTOMS—CASES.—DISEASE OF THE MITRAL AND AORTIC VALVES—SYMPTOMS—CASES.— COMPLICATIONS.—TREATMENT.

GENERAL OBSERVATIONS.—All the valves are liable to become diseased; the mitral more than the aortic. But the mitral and aortic are more liable to become affected from the inflammation travelling along the lining membrane of the heart from one to the other. The pulmonary valves are not very susceptible. Of 44 cases, in

  6 the aortic valves were alone affected.
  14 ,, mitral   ,,   ,,
  18 both the aortic and mitral valves.
  6 the pulmonary valves.

Either the aortic or mitral or both were implicated in 20 out of 45 cases of pericarditis; they were also implicated in 25 out of 54 cases of hypertrophy of the heart. The hypertrophy was generally the result of the obstruction of the valves. Exertion had a great deal to do with producing the hypertrophy.

The aortic valves often become affected in colonial fever, but as the fever subsides the alteration disappears. There is, however, a tendency left in them to become again altered on the occurrence of colonial fever or rheumatism. The mitral valve does not seem to be more liable to disease than the aortic valve, but when it becomes affected the disease is more persistent. This accounts for the greater number of cases of mitral disease which applied. It is not uncommon to see cases in which both the aortic and mitral valves are affected, but the alteration in the former disappears, leaving the latter diseased.

The aortic and mitral valves may become diseased from two causes : first, in connection with colonial fever, a disease which, like rheumatism, depends on checked perspiration ; and, secondly, on rheumatism. It is not always possible to trace or even separate the connection which exists between colonial fever and rheumatism. In both there is a tendency to produce imflammation of the lining membrane of the blood vessels ; but in the first the vessels of the bones, lungs (most frequently) and the brain and its membranes seem most susceptible ; while in the second, those of the fibrous tissues. The poison of colonial fever does not seem to be sufficiently potent to act through the blood so as to produce acute disease of the valves of the heart, but it produces congestion of the lining membrane of the vessels, and hastens the occurrence of ætheromatous deposits. In acute colonial fever the lining membrane of the heart, aorta and other vessels will be found injected. Colonial fever has ceased to be the fatal disease it was in the early days, when medical men believed more in physic—calomel and salines, with low diet and free purgation, than in keeping up the strength by nourishment and stimulants, and protecting the body from sudden variations of temperature.

Rheumatism is the great cause of disease of the aortic and mitral valves. As in pericarditis, it may travel down the aorta or its lining membrane, and affect first one valve and then the other. The pericardium may be implicated at the same time.

Influence of Sex, Age, and Occupation in predisposing to Disease of the Aortic and Mitral Valves.

The sex and age were noted in 33 cases which applied at the Institution; 18 of the number were males, and 15 females. The youngest male in which it occurred as the result of disease was 9 years of age; the youngest female, 12. The oldest males were 50 and 52; the oldest females, 55 and 65. Males seem to be more liable than females from 15 to 20 years, and after 40, as will be seen from the following table :—

		MALES.		FEMALES.
Under 15	...	1	...	1
15 to 20	...	3	...	1
20 to 30	...	4	...	5
30 to 40	...	3	...	3
40 to 50	...	5	...	3
Beyond	...	2	...	2
		18		15

Those who followed occupations which rendered them liable to be exposed to sudden changes of temperature or wet suffer more than those who were well housed and clothed. Of the 18 males, 9 were laborers, and of the 15 females, 11 were general servants, or they did washing and mangling. Females who make men's clothing, and use a heavy iron in pressing, seem also liable to suffer.

In most of the cases in which the pulmonary valves were affected, the patients were males following sedentary occupations. Violent exertion, as rowing or running, seems to excite disease of them. In one case it was caused by working a heavy chaff-cutting machine.

When hypertrophy of the heart occurred in connection with disease of the valves, the patients were either miners or they were compelled to exert themselves as laborers or as domestic servants. Of 25 cases of valvular disease which occurred in connection with hypertrophy of the heart, in 4 the aortic valves were alone affected ; in 6 the mitral alone ; and in 15 both the aortic and mitral. Alteration of the position of the aorta in hypertrophy, pericardial adhesions, and dislocation of the heart may cause an aortic murmur, and alter the mitral one so much as to render it impossible to say whether it depends on disease or not.

Influence of Pericarditis in exciting Disease of the Valves.

The pericardium may become affected simultaneously with the valves, or the inflammation may extend from one to the other. When the valvular inflammation occurs first, and the alteration in them is very marked, the friction sound of inflammation of the pericardium will be so much masked that great difficulty will be experienced in distinguishing it. The rapidity with which fluid is effused in pericarditis, and from its separating the inflamed membranes, is the cause of this difficulty.

Symptoms.—1. *Of Disease of the Aortic Valves.*

If the alteration in the valves is slight, it may escape notice ; but if considerable, and attended with pain, the sound may vary from a slight flop to a soft blowing, becoming, as the inflammation continues, more and more harsh until it assumes a rasping one.

The rasping sound is more frequently met with in cases of at least three or four weeks' duration. In the mitral valve it does not appear until the third week. Much will however depend on the acuteness of the inflammation and the rapidity with which the valve alters.

Tracing the alterations which the valves undergo in acute inflammation, it will be found that at the commencement, with pain more or less marked in the seat of the disease, fine crepitation will be heard over the valve, which will pass more or less rapidly, according to the severity of the attack, into a moist sound, and still later into a kind of flop. The first of these sounds indicates the existence of congestion of the valves; the second, of thickening; and the third, of defective action, as if they had lost some of their elasticity. In cases of some duration, with this flop there may be a distinct crackling sound, as if the valves had become hardened and parchment-like. With the occurrence of deposit on the valves, or narrowing of the opening, there will be more or less blowing sound. As the deposit on the valves, or the valves harden, the blowing sound will become rasping. I have several times had an opportunity of watching these changes. The existence of a blowing sound must not be considered as a proof that the alteration is recent. This can only be considered to be the case when there is no hypertrophy of the heart. Even in chronic cases, when the obstruction of the blood through the opening is not great, unless the patients have had to exert themselves, the hypertrophy may not be very marked. In a prostitute, aged 19, who entered the hospital, suffering from pain over the region of the aortic valves, attended with a soft blowing sound, there was no hypertrophy of the heart. She remained in for two or three weeks without the symptoms undergoing any alteration. She went out, and in less than a week re-entered, suffering from pulmonary apoplexy, and died in two or three days. The sounds on the aortic valves were not altered. On the upper borders of the aortic valves there were wart-like growths of some standing. There was no narrowing of the opening. In the next case, although the symptoms indicated that the disease was recent, the *post mortem* examination showed that it had existed for some time.

Cold, followed by difficulty of breathing and palpitation.— Pulmonary Apoplexy.—Alteration in the Aortic Valves.— Hypertrophy.—Effusion into the Pericardium.—Death.

Male, aged 27, a cab-driver, entered the hospital on the 10th of October. His health had been good until three months ago, when he began from severe cold to cough and spit up a little thick, clear phlegm. A week ago his breathing became difficult, and his heart began to palpitate and the phlegm to be tinged with blood. His face was pale; skin greasy, and warmer than natural; pulse, 120, full and hard; slight cough, with expectoration of mucous, tinged with blood and clots of blood; hands and feet œdematous. He could not lie down. He has never had rheumatism. Two years ago he had syphilis, and now there are some copper-coloured spots on the body. Urine natural, and free from albumen. General crepitation existed in both lungs, with some dulness in the upper part of the right one. He complained of slight pain in the region of the heart. The dulness in the region of the heart was six fingers' breadth deep and six across. The heart's impulse was not very much increased; it was more marked when he was made to lie on the left side. There was a loud, harsh murmer, approaching to rasping, over the aortic valves. He had never had either palpitation or difficulty of breathing, until a week ago. There was no change in the symptoms. He died on the 15th, five days after his admission.

The right arm and leg were œdematous. The pericardium contained 5 ounces of serum. Near the apex of the heart some long cellular bands passed to the pericardium. The heart was distended with blood ; the walls of the left lateral ventricle were thickened, and its lining membrane and that of the left auricle were dense. The aortic valves were thickened and cartilaginous, and the aorta throughout the thoracic region was studded with ætheromatous deposits. The mitral valve and the lining membrane of the right ventricle and auricle presented no alteration. Both the lungs were very much congested, and in the apex of the right one there were five or six apoplectic clots, varying in size from a filbert to a walnut, situated nearly close together, The lining membrane of the bronchial tubes was congested and covered with mucous, tinged with blood. The liver and kidneys were very much congested. In a case similar to this the patient recovered from the attack, but the apoplectic clots broke up, a cavity formed, and he died at the end of 14 weeks of consumption. The pericardium was adherent to the heart by a fine layer of cellular bands, not unlike wool ; the aortic valves were thickened and cartilaginous. The right lung was thickened and of a whitish-blue colour in its upper half ; its upper lobe contained an irregular cavity. Disease of the aortic valves is not liable, unless the opening is contracted, to end fatally. When it does so, it is from the occurrence of some secondary disease.

The following case will illustrate the progress of inflammation of the aorta, as observed in the colony :—Female, aged 19, a servant, three months in the colony, seen 1st of June, 1862. Twelve days ago, from exposure to cold when hot, she was seized with severe chills, followed by pains in the right and left wrist and in the right elbow. She continued to get about for another week, then from fresh cold the pain extended to the shoulders, and two days later she felt pain and uneasiness in the chest, and a harsh, long cough with a little glairy expectoration appeared.

There was but little swelling of the joints, the chief complaint being of pain at night ; the cough was not very severe ; the heat of the skin was not markedly increased, but the skin was greasy ; pulse 100, rather full ; tongue coated with white ; urine natural. There was, in addition to the sense of constriction across the chest, a feeling of uneasiness close to the left side of the sternum, between the fourth and fifth ribs, but it was so slight that it was only on drawing her attention to it after the discovery of a feeble blowing sound over the valves that she noticed it. There was no friction sound in the pericardium, and the heart's size and impulse were not increased ; there was slight respiratory harshness in the larger divisions of the bronchial tubes. Under the employment of iodide of potassium and colchicum, with nitrate of potash, and small doses of calomel and opium, the rheumatic symptoms quickly subsided, but the alteration in the valvular sounds did not disappear for four weeks, and even then the valves gave an almost parchment, flop-like sound when they came together. When first seen, the blowing sound was soft, it gradually disappeared as the system came under the influence of the mercury, and as it ceased, crepitation when the valves closed became evident. This took place about the ninth day. On the twelfth, she had a relapse from exposure to cold, and on the fourteenth the blowing sound returned, but it again disappeared in the course of a few days on giving her calomel again. She had several other attacks of rheumatism, attended by an aggravation of the heart symptoms. She died in 1864, from dropsy.

Disease of the Mitral Valve.

The mitral valve was affected alone in 14 of the cases collected; in 18 it was affected with the aortic valves. In one of the 14 cases the pulmonary valve was also affected.

CAUSES.—It is not always possible to trace the causes which excite disease of the mitral valve. It frequently exists, the patient making no complaint, and being quite unable to give any reason for its occurrence. In growing boys and girls, it can generally be traced to over-exertion; self-abuse, by weakening the strength, favors its occurrence. Rheumatism is another frequent cause, particularly of the diaphragm and pericarditis. In carditis the inflammation often extends from the aortic valves . to the mitral. (See Disease of the Aortic and Mitral Valves.) The disease of the former valve may disappear, leaving the latter affected. This seems to take place, however, but seldom. There exists, as I have before observed, a peculiarity here, namely, that when both valves are equally affected, the patients are able when the disease becomes chronic to get about, and they make but little complaint, either of palpitation or difficulty of breathing, unless they over-exert themselves or some secondary affection sets in. It is difficult to say whether pulmonary emphysema excites the mitral disease, or the cold and wet and the severe exertion to which the patients are exposed. In two of the cases the lungs were affected first. The same observations apply to drunkards. The feet are first noticed to be swollen, the liver enlarges, and the urine becomes more or less albuminous, and the pulse feeble and quick. The examination of the heart shows some alteration in the mitral valve, with increased action, difficulty of breathing, and inability to lie on the side. In the case of a girl, aged 13, the mitral disease seemed to have been excited by over-work and walking a long distance to school; the bronchitis was caused by sitting in wet clothes. She died ultimately from consumption. In a young man, suffering from tubercular consumption, the change from the occupation of a clerk, which he had followed in England, to living in a hut and being poorly clothed and fed, was the cause.*

SYMPTOMS.—There is generally more or less pain, according to the extent and acuteness of the disease, below the nipple, midway between the base and the apex of the heart. There is more constitutional disturbance than in disease of the aortic valves; the

---

* This poor fellow, like many others, had been sent out "to get rid of his consumption." It is a mistake to send consumptive patients to this colony, particularly if they are compelled to labour. From tubercular deposits being rarely observed here, the climate will be found of great service to the younger branches of scrofulous families. But it is an erroneous idea to think that there is no consumption here. It is very common; but instead of there being tubercular deposit, which softens, breaks up and forms a cavity, as in England, it commences either as pulmonary apoplexy or congestion, and then softening, cavity often forming in from seven to ten days, or bronchitis of long standing, which excites ulceration of the walls of the larger bronchial tubes and induration of the lung. Tubercles are certainly sometimes met with here, but the patients with a more or less scrofulous tendency have been exposed to the same causes which produce them in England—long exposure to cold and moisture,

face is more anxious ; the pulse is feeble and quick, ranging from
90 to 120, or even higher, and more or less irregular according to the
amount of the alteration in the valve and the extent with which it
interferes with the folds of the valve coming in opposition.   If this
is complete, the pulse will be regular, but if not the regurgitation of
the blood will render it irregular and more or less feeble.   The first
sound may vary from a soft murmur to a harsh or rasping sound.
The second sound may present but little alteration when the first
sound is soft, but when it is harsh it will be click-like or rough.
If the regurgitation is very great, the two sounds may be either
very indistinct or so mixed together that they cannot be very
easily separated, particularly when the heart's action is very much
increased.   When the anterior fold of the valve is affected, the
sound is more intense in the front, but when the posterior,
behind, along the lower border of the scapula.   In old cases, in
which there is hypertrophy of the walls, enlargement of the
liver and dropsy of the abdomen, the heart often gets twisted or
pushed to the right, so that the mitral sounds ascend up along the
aorta, and lead to the supposition that the aortic valves are
diseased.   I have alluded to a case, at page 35, in which the heart
was bent on the aorta, causing a loud murmur to be heard in it.

It is often very difficult to trace the progress of a case of disease
of this valve or the causes which produce it.   Chorea of the heart
may end by causing it.   This was particularly marked in one case,
a female, aged 48, who worked hard.   (See page 3 and *Chorea of
the Heart.)*   Patients suffering from mitral disease are very liable
to periodical attacks of palpitation.   During these attacks of palpitation the sounds will be, if the valves come in direct contact, very
much intensified.   The same will be observed during mental

excitement or bodily exertion. The exhibition of a large dose of opium or morphia, to relieve the pain, will often render the mitral sound louder, and add very much to the suffering of the patient, while digitalis and belladonna or chloroform will lessen them and give relief to the pain.

After the disease has existed some time, if the obstruction is marked, the patient being compelled to work, regurgitation will take place, and hypertrophy of the walls of the ventricle, with more or less dilatation of its cavity. The extent of the dilatation will depend on the resisting power of the walls of the ventricle. In one case, in which the walls were very thin, the cavity was twice as large as the one on the opposite side. The patient, a man 50 years of age, had been compelled to work until within three or four weeks of death.

As the disease progresses, the circulation of the blood through the lungs and liver becomes interfered with. There is either bronchitis or pulmonary emphysema. It is not always possible to trace whether these do not precede rather than follow the alteration in the valve.

COMPLICATIONS—It is not uncommon to find patients with more or less murmur in the mitral valve who make no complaint, and are able to follow their occupations, if not very laborious. I have not seen a case either of acute or chronic disease of this valve prove fatal. The acute generally ends in the chronic form, and upon this some secondary disease sets in if the patients con_ tinue to work and to be much exposed to cold and wet or to drink freely. The usual complications are in those who labour—pulmonary emphysema and enlargement of the liver, with dropsy, first of the legs, the right one usually swelling first, and then of the abdomen, and still later of the chest; but in drunkards the pulmonary symptoms, unless fluid accumulates in the chest, are usually absent. If bleeding occurs from the nose in these cases, it is oftener from the right than the left nostril. Of the 14 cases which applied to the Institution, in 8 there was disease of the lungs. In one, a female, aged 13 years, there was acute bronchitis; she died from consumption. In a male, aged 25, tubercular phthisis. In the remaining 6 cases, there was pulmonary emphysema. They were all beyond 40 years of age; 5 of the number being males and 1 female. In one of the cases, a fireman, aged 47, on board of a steamer, there was alteration of the pulmonary valves.

The next most frequent complication was paralysis. It occurred in three out of the 14 cases. In one, the patient, a hawker, aged 60, had suffered for some time with disease of the mitral valve; the face and one side became paralysed. He recovered from the paralysis. The sound in the mitral valve was harsh, but there was no regurgitation. The arcus senilis was strongly marked. In the second case, a male, aged 60, there was paralysis of the face. It disappeared. The arcus senilis was marked. In the third case, the patient, a laborer, aged 50, had suffered from "slight disease of the heart" for two years. He was suddenly seized with paralysis of one side while drinking heavily. He recovered from this, but the heart symptoms became worse. When seen, there was pain in the region of the heart, with palpitation and difficulty of breathing on exertion and on attempting to lie on the right side; pulse, 96, feeble but regular; liver a little enlarged; arcus senilis slightly marked. The size of the heart was not markedly increased. There was a loud sound like the beating of a drum heard with the first sound over the mitral valve. He was relieved by treatment; the sound continued, but it was less intense. In a medical man, less than 30 years of age, who was suffering from mitral disease of old standing, paralysis suddenly took place while out walking. He died in the course of a few months. I had not an opportunity of watching this case.

The following case presented some peculiarities, from paralysis appearing and disappearing :—

Repeated Premature Confinement and a Miscarriage.—Paralysis of one side.—Symptoms of Disease of the Heart.—Alteration in the Mitral Valve.—Softening of the Stomach.—Death.

The patient, aged 24, five years in the colony, was seen October 13th, with Dr. Stewart. She has had four premature confinements in succession, and has lately miscarried. Menstruated yesterday for a short time; the discharge was pale. She has lately had an attack of paralysis, and lost the use of one side, but from this she recovered. She has always been delicate, but never had rheumatism. The most peculiar feature of her case for some time has been extreme rapidity of the pulse, which has ranged from 120 to 140 and 150, with palpitation of the heart, and more or less difficulty of breathing on exertion, and a tendency to faint when in the erect position. She is thin and pale, eyes sunken, and surrounded with dark rings; conjunctivæ and skin tinged with yellow; liver tender and enlarged, and epigastrium tender; heart's action extended, with considerable dulness; marked bruit over the mitral valve, and a slight one over the aortic valves. She has never had rheumatism, but her father died of disease of the

F

heart. For the last two days she has vomited frequently, the matters vomited consisting of the food taken, of acid or bilious watery fluid, and thin mucous; motions and urine natural; tongue rather red at apex and sides; pulse 120, feeble; unable to sit up from the tendency to faint. She could not take any medicine from the tendency it had to excite vomiting, and for the same reason she refused to take food. She gradually got worse.

On the 15th November, the emaciation was extreme; petechial spots covered both legs; they had existed for a fortnight; pulse 120, extremely feeble; the retching was very severe, occurring every ten or fifteen minutes; the diaphragm acted very violently.

The irritability of the stomach was so extreme that everything introduced into it was rejected immediately, and was followed by severe retching; the epigastric tenderness was severe; she had not slept for some days.

Beef-tea injections were ordered, with sedative solution of opium.

She slept three or four hours after the injection of a drachm of sedative solution. She appeared to improve a little under the treatment, and although the pulse became stronger, and the tendency to fainting disappeared, and the epigastric tenderness lessened, the stomach rejected everything but a little weak brandy and water. In the night of the 19th, she had an attack of a nervous character, and the next day she was in a state of general tremulousness, which lasted three or four days.

28th.—The vomiting and retching very troublesome; no sleep for two nights; pulse, 135, very compressible, and rather vacillating; unable to be turned without danger of fainting; breathing very difficult; epigastric tenderness marked; emaciation had increased, and the dark circles round the eyes were very marked; the tremulous state had returned; the petechial spots, which had lessened in colour, had become deeper, and there was a little oozing of thin blood from the ears. The opiate injections, although the quantity of opium had been increased, had lost their power, and the beef-tea injections were returned as soon as given. On the 29th, she passed a large watery motion, mixed with a little thin blood, of the colour of port wine and water. During the day she vomited up some bile.

December 1st.—Sinking; pulse scarcely perceptible; vomited some blood during the night; bowels had acted frequently; the motions contained a considerable quantity of pale blood; she had had two slight attacks of convulsions during the night; the left arm and hand were in constant motion, but the right arm was nearly motionless; the right eyelid fell more than the left one, and the mouth was drawn a little to one side; attacks of difficulty of breathing frequent. She died in the course of the day.

The emaciation was extreme; only the posterior cerebra sinuses contained blood; the cerebrum was healthy, but the white portion appeared rather more dense and elastic than usual; the cerebellum appeared rather softer, and the grey portion rather darker than usual; about two ounces of clear serum escaped from the spinal canal; the upper part of the cord was healthy; the liver was large and slightly cirrhosed; the lungs healthy; the heart was twice its natural size; the wall of the left ventricle was one inch in thickness; the mitral valve was dense and cartilaginous, and rough from the presence of warty growths; the aortic valves were a little denser than usual; the right side of the heart was distended with blood, but the left was empty; the womb and ovaries were small, but the lining membrane of the former was of a deep venous hue on its posterior part; the stomach contained about half a pint of bilious fluid tinged with blood, of the same colour as that passed from the bowels; its mucous membrane was of a slate colour; it could be easily raised in flakes of the size of a sixpence, and removed by scraping with the edge of the scalpel.

The occurrence of the paralysis and the nervous symptoms would almost lead the practitioner to think that there was considerable alteration going on in the brain. It is possible that she had drank a good deal.

In another case, the patient, a man about 30, died from apoplexy. The alteration in the mitral valve was not very marked ; he had drank heavily, smoked to excess, and was much addicted to women. He suffered very much from chorea of the heart for some time after his first sleep.

In two of the other 14 cases, amenorrhœa existed. In one, a a servant aged 26, there was chlorosis; the liver was enlarged, and the legs were very much swollen. The mitral murmur was harsh, and although she recovered, the sound remained unaltered. In the other case, the menstrual discharge had been stopped for several months, and then dropsy set in. She was tapped several times, and from 1½ to 2 gallons of fluid removed. She died suddenly, after an attack (the second within a few days) of profuse menorrhagæ. The heart was enlarged; the mitral valve was very dense, the opening was narrowed. The pericardium adhered to the heart. She had had a severe attack of rheumatic fever two years before the dropsy set in, from exposure to cold and wet soon after her confinement.

In another of the fourteen cases, the patient, a cooper, had drunk very heavily for some years. The mitral murmur was not much altered; the liver was very much enlarged, legs dropsical, and urine contained albumen. He recovered, and while he abstained from drink remained well, there being but little alteration in the mitral sounds. The liver diminished in size, and the albumen disappeared. If he drank, the heart symptoms became aggravated, the mitral murmur increased in loudness, and pain was excited in the region of the heart, and palpitation and inability to lie on the side, and the liver began to enlarge. In one of the patients, eating poisonous fish brought on an attack of palpitation of the heart, and aggravated the mitral alteration. The heart did not recover its usual state for several days. In a professional man, about 45 years of age, suffering in a similar manner, from being unable to abstain from drink, death ensued. This is the usual result when the patient continues to drink. In another of the 14 cases, the patient, a male aged 14 years, was seen June the 24th, suffering with pain in the region of the heart, palpitation and cough, with a little thin mucous expectoration. He had had rheumatism for 14 days; six months ago, he also had an attack; now the legs are affected, but in the former attack, he had pain in the lower part of the chest, and the "heart beat a good deal, but it got better." Now the palpitation has been severe, since the commencement of the attack, which was excited by exposure to cold and wet. His face is pale; skin natural; urine high coloured; liver large, extending two fingers breadth, below the ribs; pulse 118, rather feeble. The heart's dulness is increased in extent, and there is a loud harsh murmur heard in front, over the region of the mitral valve, and along the lower border of the scapula. A slight clicking sound existed over the aortic valves. By the exhibition of digitalis with iodide of potash, small doses of calomel and opium at night, for a few nights, and blistering, the palpitation and the pain disappeared, the liver diminished in size, and the pulse sank to 90. On the 2nd August, although able to walk without much difficulty, there was still a harsh sound over the mitral valve, but the click-like sound had disappeared from the aortic valve. He recovered so far as to be able to take a situation, walk at a moderate pace without difficulty; but the murmur, although much less harsh, continued.

Since the table of the 14 cases was compiled, 4 other cases have come under my notice. In one, a lad aged fifteen years, from working a chaff-cutting machine; a second, aged 14, working as a lithographic printer; a third, a digger aged 20, from wheeling heavily loaded barrows up an inclined plane. He had not been accustomed to this kind of work. A fourth, a servant aged 28; she had had rheumatic fever 18 months before. Her menstrual discharge had stopped 8 months. The three first cases recovered by strict rest; the fourth was relieved, and her menstrual discharge returned.

Disease of the Aortic and Mitral Valves.

The aortic and mitral valves were affected in 18 cases. It is met with both in the acute and chronic forms. They are rarely alone affected, pericarditis being common in the acute, and hypertrophy in the chronic cases. The disease, particularly when it occurs in connection with rheumatic fever, is acute at the commencement; when the fever disappears, it passes into the chronic form; the patients are able to go about with more or less ease, according to the amount of obstruction. If this is equal in both valves, it is surprising to see how the heart adapts itself to the alteration. A patient may go in this state for years, as will be seen by the cases reported. Another attack, from exposure to wet and cold, may then occur, and the valves become more altered. They usually recover again for a time, unless acute congestion of the lungs, liver, and kidneys is excited. If the congestion of the lungs is very severe, or neglected, the case may end fatally, particularly if there is suppression of urine.

Causes of the Acute Form.

Rheumatism is the most frequent cause of the acute form. This was observed in 6 out of 7 cases. In three of these cases, the disease ended fatally from congestion, or congestion and apoplexy of both lungs. In the seventh case, the disease was excited by lying out exposed to cold and wet; there was complete suppression of urine; the brain and its membranes were altered, the liver, kidneys, and lungs were congested, the last contained apoplectic clots. The lining membrane of the heart, and larger vessels, and the valves were altered. In 5 of the 6 rheumatic cases, the patients had previously had acute rheumatism, and the aggravation of the mitral and aortic disease was due either to another attack or to congestion of the lungs, which in the three cases that ended fatally was severe enough alone to cause death. In one case, the patient, a female aged 17, four months in the colony, had had repeated attacks of rheumatism, but it was the first time she had complained of pain in the region of the heart. In a second case, the patient, a male aged 20, three months in the colony, had had rheumatism in England five years before. This was the first time that he had complained of his heart. In the third case, the patient, a female aged 21 (a native), had had rheumatism 2 years before; her heart was then affected. In a fourth case, the patient, a male aged 28, seven years in the colony, had had rheumatism soon after he landed. His heart was affected, and he had suffered from palpitation and difficulty of breathing, if he exerted himself very much. In the fifth case, the patient, a digger aged 40, seventeen years in the colony, had been discharged from the artillery for disease of the heart. He had had rheumatism 5 years before, but his heart, he said, was not then affected, but traces of old pericarditis were found after death.

Repeated Attacks of Rheumatism in England.—Rheumatism of the Shoulders and Ankles, with acute Disease of the Aortic and Mitral Valves.—Relieved.

A servant, aged 17, four months in the colony, entered the hospital on the 23rd February. She had been ill 7 days with acute rheumatism of the shoulders and ankles. The attack was brought on from sudden exposure to cold when very warm. She has had several attacks of rheumatism in England, the first when she was 9 years of age, but this is the first time she has ever had pain in the region of the heart, and palpitation. The heart became affected three days before she was admitted. The rheumatism was not very strongly marked; there was not much swelling of the joints; pulse 108; tongue white; skin moist; the perspiration acid to the taste, and the urine loaded with lithiates. She was delicate looking; had never menstruated; but for three years had suffered from leucorrhœa. There was a rather loud blowing murmur, both at the apex and the base of the heart, with the first sound; the second sound in both situations was flop-like. There was considerable congestion of the bronchial mucous membrane, with harsh cough and expectoration of thick tenacious mucous, tinged with blood. In the course of the next 10 days, the first sound over both valves became more and more rasping, and the second sound parchment-like. The cough and expectoration disappeared. The treatment was not very active. She remained in the hospital until the 10th June. There was but little alteration in the heart sounds. She was able to walk about without difficulty of breathing or palpitation. The heart was not increased in size. She had not menstruated.

The next case bears some resemblance to this one. The patient, a male aged 20, a servant, three months in the colony, entered the hospital August 20th, suffering from pains all over the body, and in the diaphragm, of 5 days' duration, caused by exposure to wet and cold. There was a loud murmur both at the base and apex of the heart with the first sound, the second is click-like; pulse 120; skin rather dry; urine high-coloured, loaded with lithiates; tongue white. Calomel and opium, with salines ordered. Had rheumatic fever five years before in England. His joints were then affected and much swollen. But now it is the muscles, the joints being but very slightly implicated. His heart he says was not before affected. September 2nd.—The murmur is broken up, and mixed with a soft click-like sound; pulse 90; the pains in the limbs nearly gone; tongue clean, and urine clear, but high-coloured. Mouth slightly affected by the calomel. The pulse by the 11th had sunk to 66. He went out at the end of the month. The heart sounds were still a little altered.

In the third case, the patient recovered, but there was permanent alteration of the valves. The patient, aged 21, a native, the mother of a child residing in a badly-constructed house, seen October 16th, suffering from pains in the whole of the body, with chills and heats. Two or three days before, she had got cold and wet while washing; pulse quick, 108; tongue coated with white fur; urine high-coloured; sweat acid. She had had rheumatism 2 years before. A brother died last year from inflammation of the left lung and pericarditis, after a few days' illness. There was no alterations of the heart's sounds.

19th.—Began to complain last night of pain in the region of the heart; pulse 116, regular. There is marked alterations in the mitral and aortic valves with the first sound; rheumatism the same, but there is no swelling of the joints. She had been exposed to cold the day before.

On the 26th, by blistering the chest, and the exhibition of iodide of potash, with opium, colchicum, and bicarbonate of potash, she was very much better.

On the 29th, from fresh exposure to cold and wet, she was very much worse, and complained of severe pain in the shoulders and diaphragm.

30th.—Pulse 120, regular; face anxious, and eyebrows knitted; skin dry and harsh; tongue white. The murmurs in the valves were very much louder. The second sound in both valves rather floppish. Complains of considerable pain in the region of the heart. Eight leeches were applied, followed by a warm linseed meal poultice, and ½-grain doses of calomel, with half-a-grain of opium every four hours, in addition to the iodide of potash mixture. By the 6th, she had greatly improved. The mercury had not affected her gums; urine clear; sweat not acid. The aortic sounds were more natural, but the first sound in the mitral valve was as harsh; this had gradually set in, although the rheumatic symptoms had disappeared, and the pulse had sunk to 108.

She was not again seen until the 18th of January, when she applied to the institution for medicine to bring on the menstrual discharge. The heart's impulse was considerable; pulse 114, feeble, but regular; her feet have a tendency to swell; urine natural. The first sound in both the aortic and mitral valves was loud and harsh. By the exhibition of minute doses of bichloride of mercury with iron, the sounds in the valves improved and the menstrual discharge re-appeared.

In the following cases death ensued :—

In the first case, the patient aged 22, a prostitute, entered the hospital on the 23rd of July. Ten days before, from lying out while drunk, she was seized with sore throat, chills and heats, with pains of a rheumatic nature in the limbs, but she had not had rheumatism before. There is a loud murmur over the aortic and mitral valves with the first sound; the second sound was slightly altered. A blister was applied to the heart, and iodide of potash ordered. She was taken out at the end of a few days, but was brought back within a week, in a dying state.

Both lungs were very much congested, and contained clots of blood. The upper aspect of the aortic valves was covered with warty growths. The valves were rather thickened, but not otherwise altered. There were similar growths on the upper aspect of the folds of the mitral valve. The pulmonary valves were thickened, and the lining membrane of the right side of the heart and the pulmonary artery was deeply congested.

In the next case, the patient, aged 28, a shoemaker,

seven years in the colony, seen with Dr. Stewart on the 2nd of August. He had had rheumatism soon after he landed. He had severe pain in the chest with the attack, and since has suffered from palpitation and difficulty of breathing on exertion. From exposure to cold and wet seven days back, his breathing became difficult, with fever, cough, and expectoration of clots of blood, and of mucous tinged with blood. His face was very pale and anxious, pulse very feeble and irregular, the legs were dropsical, and the abdomen contained fluid. The lungs were very much congested, and there was ægophony in the posterior part of the chest. The difficulty of breathing was extreme; he sat up in bed gasping for breath. At times he fell into a state of syncope, the pulse ceasing to beat. The heart's action was very irregular, beating tumultuous for eight seconds and then ceasing for two or three. The mitral and aortic sounds were very much confused; the valves seemed to act incompletely. There was a loud murmur heard over the pulmonary valves. The symptoms underwent but little alteration during the next two days. He spat up a considerable quantity of blood. His face and hands became œdematous. He died suddenly while attempting to get out of bed. There was dropsy of the legs, walls of the abdomen, back, and hands. Both pleural cavities contained from four to six ounces of red-colored serum. The lungs were very much congested; the apex of the right one contained clots of blood, which had broken up the tissue of the lung. The pericardium adhered to the heart by old adhesions. The heart was enlarged; the walls of the left ventricle thickened; the substance of the heart was very friable. The lining membrane of the heart and large vessels was dark-colored. The aortic and mitral valves were thickened and covered with deposits; the tricuspid and pulmonary valves were thickened. The liver was large and congested; the kidneys the same. The abdominal cavity contained several pints of fluid. In the next case, the disease was more chronic :—

Symptoms of Disease of the Aortic and Mitral Valves.— Congestion of the Lungs.—Dropsy.—Death.

A digger, aged 40, seventeen years in the colony, entered the hospital October the 2nd, suffering for ten or twelve weeks with pain in the region of the heart shooting through to back, with palpitation and difficulty of breathing on exertion, and inability to lie on the left side. The attack commenced while working in wet ground, and was attended with pains all over the body, heats and chills, high-colored urine, cough, and expectoration of mucous. Had an attack of rheumatism five years ago; was ill for nine months, but does not think that his heart was affected, for he had no pain or palpitation. He left the artillery when 21 for disease of the heart. When admitted his pulse was 108, rather full; tongue white; rheumatic pains better than they had been; still pain in the region of the heart; loud murmur over the mitral and aortic valves; with the second murmur there was a flop-like sound. Both valves were equally affected. There was some bronchial irritation, with cough, and the expectoration of mucous tinged with blood. The base of the heart encroached on the right side of the sternum. The heart's dulness was six fingers' breath from base to apex, and five fingers broad at the base. November 8th.— There has been but little alteration in the symptoms. He has spat blood nearly every morning from the night being cold. Thinks he has got cold. To-day his face is pale and anxious; breathing difficult, compelling him to sit up in bed; pulse 96, full and strong. There was considerable congestion of the posterior and lower lobes of the left lung and in the apex of the right one. There was some ægophony in the post part of the chest. The liver dulness extends below the ribs. November 11th.—The thighs and legs are beginning to swell. By the 15th, fluid had begun to accumulate in the abdomen; the legs were more swollen; urine scanty, free from albumen; pulse 102, rather wiry. November 27th.—The dropsy has gradually increased; he sweats profusely; coughs a good deal, and brings up a little white phlegm. Is unable to lie on the right side; pulse 108, full; passes about 30 ounces of urine in the 24 hours. He died suddenly on the 30th, after bringing up about half-a-pint of blood.

The lungs were very much congested; there was fluid in both pleural cavities. There were several pints of fluid in the abdomen. The heart was displaced towards the right side. It was considerably enlarged; the aortic valves were very dense and hard; the mitral valve the same, but not to the same extent. There were traces of old pericarditis.

In the next case, the symptoms appeared to be the result of uræmic poisoning :—A man, about twenty-five years of age, was brought into the hospital in an insensible state. He could be roused, but no history of his case could be obtained. He had been found lying out in the cold and wet. His pupils were dilated, breathing hurried, pulse full and hard, arms rigid and contracted, and when an attempt was made to extend them he uttered low cries. Pressure over the region of the kidneys also excited cries. The heart's action was very much increased, and there was a distinct, though not a very loud murmur heard over the mitral and aortic valves. The case seemed to be one of uræmic poison, from there being no urine in the bladder. Mucous râles existed in the course of the large bronchial tubes, and crepitation in the lungs posteriorly. He died 48 hours after admission, without any very marked alteration in the symptoms.

The arachnoid membrane was opaque, and there were several drams of serum in its cavity. The vessels on the surface of the brain were congested, and the brain contained more blood points than usual. The right lung was congested, and in the centre of its apex the tissue was very soft and intermingled with clots of blood. The left lung was also congested and adherent to the pericardium. The bronchial tubes of both lungs contained mucous tinged with blood. The pericardium contained several ounces of red-colored serum, and its lining was injected. The heart was distended, with dark clots. The aortic and mitral valves were thickened and covered with fibrinous deposits. The lining membrane of the heart, aorta, and pulmonary arteries were deeply congested. The walls of the heart were congested. The liver was gorged with blood and enlarged; the kidneys the same; the bladder was empty. This case bore a strong resemblance to the acute arteritis sometimes met with in England, and to the pericarditis, complicated with endocarditis, which is sometimes met with here after scarlet fever, and occasionally after measles. (See page 40.)

Chronic Disease of the Aortic and Mitral Valves.

Chronic disease of the aortic and mitral valves is generally the result of an acute attack of a more or less marked character. It is often very difficult to trace when the first attack occurred, from its having been so slight as to give the patient but little anxiety.

A second usually occurs, and generally of a severer character, and produces palpitation and difficulty of breathing on exertion, and pain in the region of the heart. This form does not generally prove fatal. From exertion hypertrophy of the heart will be excited, and occasionally dilatation of the aorta; from exposure to cold and wet the lungs may become congested, or bronchitis, asthma, pulmonary apoplexy, or the form of phthisis observed in the colony produced, or the liver may become congested—a not unfrequent complication when the lungs are similarly affected, and jaundice, more or less marked, produced. In drunkards disease of the kidneys may then occur; the urine becomes more or less albuminous, and dropsy occurs. These cases are nearly always accompanied by enlargement of the liver, and often in males with hæmorrhage, from the bowels of a persistent character, or bilious diarrhœa, or frequent and severe attacks of bleeding from the nose. Females during the menstrual period are not liable to these hæmorrhagic attacks, but the menstrual discharge is often unusually profuse. In one case, the patient had ceased to menstruate for two years. She had constant diarrhœa; she took scarcely any food, living nearly altogether on drink during the last six or eight months of life. Her heart was empty, it was in a state of fatty degeneration, but not enlarged; the liver was fatty, it was not very much enlarged. The aortic and mitral valves were cartilaginous. The valvular disease had been of long standing. There was considerable deposit of fat under the skin, and about the viscera. She took no exercise during the last 18 months of life. Her pulse was scarcely ever under 130, and very feeble. She died suddenly, while suffering more severely than usual from diarrhœa.

The following cases are the best illustrations of this form of the disease I have been able to collect :—

Palpitation early in Life, six months before admission into the hospital, symptoms of Valvular Disease.—Slight Jaundice, and Congestion of the Lungs.

Male, aged 43, harness-maker, three years in the colony, entered the hospital June 14th.   He states that in January, from exposure to cold and wet, he was seized with difficulty of breathing and palpitation of the heart, was under homœopathic treatment, but received no benefit from it.   Turkish baths were then ordered; they relieved him very much.   His health has always been good. When 18, from over exertion, he suffered severely from palpitation, but by rest it disappeared at the end of 12 months.

On admission, he presented the following state :—Skin and conjunctiva slightly yellow ; liver slightly enlarged and tender ; no dropsy ; urine dark coloured.   Mucous wheese existed in both lungs, with expectoration of mucous of a yellow tinge, streaked with blood, and bitter tasted, of from three to four ounces in the 24 hours.   The heart's dulness was not much increased, but there was a blowing murmur over the mitral and aortic valves with the first sound, and a flop with the second.   There was rather more alteration in the mitral than in the aortic valves.   The mitral valve did not act so freely as the aortic valves.   The sounds were well marked along the lower border of the scapula.

July 12th.—There was but little alteration in the symptoms referable to the valves.   The jaundice disappeared, and the lung symptoms improved.   He left the hospital soon after.   This case bears a strong resemblance to several others reported.   From being better housed and cared for, the symptoms did not assume the same severity. In the next case, the liver was the organ most implicated.

Chronic Disease of the Aortic and Mitral Valves.—Attack of Congestion of the Liver, with an aggravation of the Heart Symptoms, followed by Dropsy.

A butcher, aged 30, seven years in the colony, entered the hospital on September the 30th. He has been ill 8 months. While working in the wet he began to suffer from flying pains in the body limbs, and chest. He had chills and heats but was not ill enough to be confined to bed. One day soon after this while exerting himself he was seized with palpitation of the heart and difficulty of breathing. The palpitation and difficulty of breathing has continued, and prevented him following his business. He was in the hospital 10 weeks ago, suffering from symptoms of congestion of the liver, with slight jaundice, and an aggravation of the heart symptoms. He left at the end of a fortnight, slightly relieved, and went into the country. He has got gradually worse, and now his feet, legs, and abdomen are very dropsical. The dropsy set in immediately after he left the hospital; the right leg began to swell first. His face is congested; neck short; has never drank hard. Heart's dulness from base to apex 6½ fingers' breadth. Liver dulness commences 1 finger's breadth below the nipple, and extends to below the edge of the ribs. The heart's action is increased, and there is a loud, harsh murmur in the mitral and aortic valves. Pulse 108, regular; urine dark coloured, free from albumen, from 1 to 2 pints being passed in the course of the 24 hours; bowels act freely, the motions, which were a few weeks ago green, are now of a bright yellow colour; tongue clean and moist, but the conjunctivæ and the skin, are slightly yellow.

Oct. 16.—Has increased very much in size, and is unable to lie down; pulse 120, feeble. He was tapped, and 92 ounces of fluid withdrawn. He was in the hospital for some time. He was tapped several times with but temporary relief. He went out, and died in March. This is a good illustration of a class of cases often met with here. They seldom recover if of any duration, but remain in the same state for several months, dying sometimes suddenly, or from some secondary affection, such as congestion of the lungs, vomiting, or diarrhœa, or both, or from exhaustion in consequence of the extreme heat.

In the next case the lungs were more implicated than any other organ.

Hooping Cough, followed by symptoms of Disease of Valves, four years before entrance into hospital.—Cold, followed by symptoms of Broncho-pneumonia and Cavity in the Lung.

A domestic, aged 20, thin and pale, eight months in the colony, entered the hospital on June 11th. She states that she has been ill 4 years. She was first taken with hooping-cough, which left her with a short dry cough, difficulty of breathing, and palpitation of the heart on exertion. She usually spat up about two table-spoonsful of frothy phlegm in the 24 hours ; when the cough was severe it was sometimes mixed with blood.

A month ago from exposure, she was seized with cold chills ; the cough got worse, and the phlegm increased in quantity, but it was free from blood. She ceased to menstruate during the

voyage out,* but the last three months it has returned; it
is regular, and of its natural colour. Her tongue is white;
thirst severe; skin hot, particularly at night; pupils large; pulse
120 ; she has been unable to lie down for several nights. She
has lost flesh very rapidly of late; pulse 120, rather feeble; the
phlegm is copious, yellow, and mixed with air. Loud râles can
be heard in both lungs, with more or less coarse crepitation.
There is considerable dulness in the region of the left nipple.
The heart's action is increased, and there is a harsh murmur over
the aortic valves with the first sound, and over the mitral valve,
but it is rather softer. There is no dropsy. The urine is high
coloured. The liver dulness is considerable. She lies on the left
side; if she attempts to lie on the right, cough is excited.

The symptoms underwent but little alteration during the next
month.

July 8th.—There seemed to be some tendency to the formation
of cavity in the apex of the right lung. The cough was very
severe in the morning, it was spasmodic, almost like hooping-
cough. The phlegm varied, consisting when the cough was
severe of frothy mucous, but at other times it was yellow and
heavy. She left the hospital soon after this, and was lost sight of.

Displacement of the heart is sometimes observed. This occurred
in a case like the last one, in the hospital. The woman, a servant
25 years of age, said that three days before admission, the heart
was felt for the first time to beat on the right side. The heart was
so covered by the lung that its exact position could not be determined. The aorta was bent soon after it sprang from the heart.
There was a peculiar, loud, harsh whiss, like that produced in sharp-

* It is common for the menstrual discharge to stop while at sea.

ening a scythe, heard along it. She could not lie on the left side, the usual position chosen, but on the right. She had been ill for two years with difficulty of breathing, palpitation of the heart, cough, and expectoration of mucous, which was sometimes thin, frothy, and green, at others glairy, like the white of egg. There were coarse râles in both lungs, in the course of the bronchial tubes, and more or less coarse crepitation in the posterior part of both lungs. There was no dulness on percussion in the pericardial region. This part was very clear and resonant, and fine crepitation could be heard. I had not an opportunity of watching the case after the first week.

In two other cases in which the heart was displaced, the patients were advanced in pregnancy. In one case which I saw with my late friend, Dr. Stewart, on the 10th July, 1867, symptoms of acute congestion of the lungs had set in suddenly, from exposure to cold and wet while washing. The patient, 36 years of age, had had disease of the heart for four or five years, but she had been able to get about with comparative comfort.

On the third day, she presented the following state :—Unable to lie down; respiration very difficult and quick; face congested, expectoration scanty, and streaked with blood. Loud râles and coarse crepitation existed in the anterior part of both lungs, the posterior part could not be examined ; the pulse was 130, full and strong ; no mitral murmur could be discovered, but there was a loud, harsh one heard over the aortic valves, and much higher than usual. The arch of the aorta could be easily felt behind the sternum. Her breasts were very large, and she was very stout. The border of the uterus was very high up. Twelve leeches were applied to the chest, with relief to the breathing, but they reduced the strength. The membranes were ruptured, and the liquor amni allowed to drain off. A large quantity escaped, and the womb descended lower down. This gave her considerable relief, and at the end of 36 hours, after several doses of ergot, labour set in, and she was delivered of an eight months' dead child. The delivery was followed by a marked improvement in the lung symptoms. She recovered. When seen six months later, she was able to get about; the alteration in the aortic and mitral valves was so nearly equal that she could walk, and do light house work without difficulty. She is I believe still living.

In the next case, labour did not set in for some time after the membranes were ruptured. The patient, aged 28, was seen, with Drs. Graham and Martin, January 23rd, 1862. She had had rheumatic fever in England, in 1856, and was then ill for four months. Her heart was then slightly affected. She came to the colony in 1857, and has had no rheumatism or any other disease since. She is now seven months advanced in pregnancy. Three months ago, she began to suffer from more palpitation of the heart than she had done since in the colony, and her breathing got difficult. She has been confined to bed 14 days. There is an alteration in the position of the heart; its base lies close to the right side of the sternum, on a level with the cartilage of the fourth rib; its apex to the left, and just opposite the nipple. The arch of the aorta can be easily felt behind the first bone of the sternum. There is a loud murmur over the mitral and aortic valves, and considerable congestion of the posterior part of both lungs, but no ægophony. The pulse is 130, hard and jerking. The abdomen does not seem to be very large, although the border of the uterus was considerably above the navel. There was no dropsy of the legs.

It was thought advisable to rupture the membranes. This was done. A large quantity of liquor amni drained away until the 12th of February, when she gave birth to a very small child.*

Dilatation of the aorta is a complication sometimes met with. If the pulmonary artery is pressed upon, it will also be dilated. The former depends on weakness of the walls of the aorta, and is not confined to disease of the aortic and mitral valves, but is observed in hypertrophy of the heart, both with and without disease of the aortic valves.

Palpitation from over-exertion, later from Exposure, Pain in the region of the Heart, and a return of the Palpitation, with Difficulty of Breathing.—Disease of the Valves, and Dilatation of the Aorta and Pulmonary Artery.

A married domestic, aged 28 years, 4 years in the colony, entered the hospital July 12th. Five or six months ago, after a hard day's washing, she was taken with palpitation of the heart, which lasted a few days. She continued pretty well up to a month ago, when the palpitation returned. It was accompanied by pain near the left breast, and difficulty of breathing. She thinks she must have got cold, for she always felt chilly, but had no distinct attack of fever, and was very weak. A blister was applied at the end of a fortnight. It relieved the pain in the chest. Thinks her disease was caused by sleeping in a damp room. On admission she presented the following state :—Face

---

* I have several times observed that digitalis, when given for some time in pregnancy, will check the growth of the child. I once gave it to a woman whose pelvis was very narrow. The child was born without the aid of instruments ; it was very small, but lived.

pale; skin dry, harsh and cool; cough with expectoration of frothy mucus, tinged occasionally with blood ; unable to walk from the palpitation of the heart, and difficulty of breathing, across the room. She has never had rheumatism. Her menstrual discharge has been regular, and of a good colour, until the last time, when it was pale and scanty. The pulse in the left wrist is 84, irregular, the 6th or 8th beat stopping for at least a second, but in the right one, it is 72, very oscillating and quite distinct from the pulse in the other. The heart measured 5½ fingers' breadth from the base to the apex, and the same transversely. There was a loud click with a blowing murmur at the apex of the heart with the first sound, it extended to the base, and up along the aorta. It could be heard along the lower border of the scapula. There was no well-defined second sound. In the pulmonary artery there was a slight increase in its impulse, and where its right division passed behind the aorta there was a confused sound, which commenced suddenly and extended for a short distance up the aorta. The external jugular veins were enlarged. The pulsations in the left carotid artery were 66 in the minute, and very feeble ; in the right one they were also very feeble. The arch of the aorta laid high behind the first bone of the sternum. She says that she can lie on either side without difficulty. In the right lung, the respiration is harsh, and on coughing, the bronchial sound is rasping, but in the left it is feeble, as if the lung was pressed upon. She expectorates about three-fourths of a cupful of mucus in the 24 hours.

August 8th.—She has been getting gradually worse; the breathing is very difficult; she feels weaker. For a week she has had water brush, and brought up saltish-tasted but clear fluid.

She cannot lie so well on the left side as she did. The menstrual discharge is paler and scantier than before.

Sep. 10th.—Has improved lately. Is able to walk about a little, but very slowly. The pulse is more equal, on the right side it is 66, and on the left 60. The aorta is not so easily felt. The other symptoms remain the same. The improvement is due to digitalis and iron and strict rest.

Oct. 1st.—Has not menstruated. She has not had the iron and digitalis lately. The palpitation is now very troublesome, but on returning to the iron and digitalis it was relieved.

This patient was lost sight of.

In a similar case, in which the patient died in the hospital, while suffering from similar symptoms, the heart was enlarged, the mitral and aortic valves were cartilaginous, and did not close completely. The aorta was twice the usual size; its walls were much thinner than usual. The dilatation was more marked from its commencement to the arch, than in the other part of its course. The pulmonary artery was so large that it readily admitted three of the fingers. The right ventricle was enlarged, but its walls were not thickened.

Albuminuria is another complication met with in those who have drank heavily.

A man about 25 years of age was brought into the hospital in a semi-comatose state, suffering from dropsy of the legs and abdomen. His liver was enlarged; skin pallid, with a slightly jaundiced tinge; urine scanty but albuminous. His pulse was irregular and feeble, with regurgitation in both aortic and mitral valves. Nothing could be gleaned of his previous history, beyond that he had had rheumatism 6 years before in the colony, with spitting of blood, and pain in the region of the heart, with palpitation. He had evidently spent all he could get in drink. The honorary physician ordered him a dose of podophylline. The only effect it had was to aggravate the jaundice, and increase the coma. He died in this state. The body was generally jaundiced; the chest contained fluid, and the lungs masses of blood, varying in size from a filbert to a walnut, intermixed with the tissue of the lung, which was very soft, and easily broken up. The right lung was adherent to the walls of the chest; the heart was enlarged; the aortic valves ossified, and adherent to each other; the mitral valve was in a similar state; the liver was enlarged and congested; the kidneys the same, and in parts fatty; the gall bladder was empty.

I lately saw, with a medical friend, a man about 40 years of age, suffering similarly. His face and skin were of a yellow waxy hue; lips and gums pallid. He had suffered from repeated attacks of diarrhœa and hæmorrhage from the bowels. During the last few days his breathing had become very difficult, from some obstruction about the glottis. There was some slight swelling on the right side of the root of the tongue. He was unable to swallow. Any attempt to examine the throat caused very severe pain, and aggravated the difficulty in breathing. The liver was enlarged, and there was some tenderness over the region of the kidneys. The lungs were congested in their posterior parts; the legs considerably swollen, but no fluid could be found in the abdomen. There was a murmur over the mitral and aortic valves; pulse 84, full; pupils were not markedly dilated. He had not passed any urine lately. He was very restless. Blistering fluid was applied to the throat. A little vesication was caused, it seemed to relieve the throat. The next morning he was still restless; pulse rather feebler; did not answer questions so readily as before; his aspect was dull and heavy. He got rapidly worse about the middle of the day, and died in the evening. He had evidently drank heavily. His health had been bad for some time.

TREATMENT.—It is very difficult to say what plan of treatment, from every case presenting some peculiarity, is likely to benefit. Strict rest is necessary, and avoiding everything that can aggravate the palpitation or the lung disease. Smoking to excess; drinking heavily of spirits, and over exertion, with exposure to cold and wet, or living in a damp house, are sure to increase the disease, and take away the chances of recovery. I have never seen a drunkard, and very seldom a labourer or a domestic recover, if they continued to be exposed to the causes which produced the disease. In the acute forms of the disease, leeches, or even bleeding from the arm, if the pulse and the general state of the health will bear it, with calomel and opium given until the mouth is rendered sore, with iodide of potash, alkalies, digitalis, and tincture of henbane, will be found of service. Blisters will assist very much in removing pain.

When the lungs become congested, leeching and blistering will be found of service, if used early, and before the lung breaks up. Opium is not borne well in these cases. Turpentine, in doses of from 10 to 20 minims, every two, three, or four hours, will often act very beneficially, particularly when the congestion is of a low type. In a case attended by Dr. Gregory and myself, it cleared the lungs in the course of a few hours, caused copious expectoration, and profuse perspirations, steadied the pulse, and reduced it from 150 to 120.

When the liver was enlarged, blisters, with bromide and iodide of potash, digitalis, with ammonia and jalap, in small doses, were of great service. The dropsy of the legs may be sometimes kept down by soaking them in warm water every night. Small blisters to the calves of the legs, keeping them open with savin ointment, are often useful. Care must be taken not to apply very large ones, or low down on the leg, as they are apt to degenerate into troublesome sores. Scarifying the calves with a small scarificator, or puncturing them with the German needles, and then applying a small cupping glass, will often draw off several ounces of fluid. This can be repeated every second or third day, if necessary. Patients do not seem to bear large doses of digitalis so well here as in England. When given, it should be combined with ammonia. Infusion of broom with juniper berries, and supertartrate of potash, are often useful. Packing the patient in a *warm, wet* sheet, and placing bottles filled with hot water about the sides and legs, will, by causing copious perspiration, relieve the liver and kidneys.

In chronic disease of the valves, I have found small doses of bichloride of mercury, 1-24th of a grain, with from three to five drops of the solution of perchloride of iron, and five of tincture of digitalis, of great benefit. It will generally restore the menstrual discharge, which is often suppressed here, and cause the deposits about the valves, if not of very long standing, to be absorbed. Its value was well shown in a case reported a few pages back. It lessened the alterations in the valves, and caused the menstrual discharge to reappear. Its efficacy was also great in a case of disease of the mitral valve, in which the menstrual discharge had been suppressed for several months.

CHAPTER VI.

DISEASES OF THE RIGHT SIDE OF THE HEART. —
(ENDOCARDITIS).

DISEASES OF THE PULMONARY ARTERY AND VALVES.

THE right side of the heart is not nearly so liable to disease as the left. It is more often affected however than is generally supposed, but from the symptoms being more obscure, and referable to the lungs or the left side of the heart, the existence of disease in it is overlooked. The father of pathological anatomy, Morgagni,* speaks of the rarity of disease of the pulmonary artery. Laennec,† Albers,‡ and nearly every writer on diseases of the heart have re-echoed this statement. Dr. Copeland, the most learned medical writer that England has ever produced, does not devote in his invaluable Dictionary a chapter to its diseases. It is just possible that his silence may have caused the reticence usually observed in those men who *make* books out of other men's labours.§ Hope ‖ only found in 1000 cases which he examined

* De Sedibus et Caus : Morb.
† De L'auscultation, Mèdiate.
‡ Anatomie Pathol : Brust. Libr. II., Sec. viii.
§ A very "distinguished" fellow of a Scotch Royal College, *in* Melbourne does still better. It may be said of him :—

> " Quâ factus ratione sit, requiris,
> Qui nunquam futuit, pater Jacobus?
> H ——————— dicat, istud,
> Qui scribit nihil, tamen auctor est."

He may be solaced by—

> " Carmina Jacobus emit : recitat sua carmina Jacobus."
> " Nam, quod emas, possis dicere jure tuum."

‖ Diseases of the heart, 3rd edition, 225.

calcarious deposits in the coats of the pulmonary artery in one, and cartilaginous and steatomatous deposits, and dilatation, in three or four. Of 310 cases which applied for relief at the institution, only six presented alterations in the pulmonary valves and artery.

The pulmonary artery and its valves are liable to—

1. Spasm of the valvular opening.

2. Inflammation of the lining membrane of the artery and its valves, with or without the formation of fibrinous clots.

3. Dilatation of the pulmonary artery.

4. Deposits or thickening of the valves, and deposits in the artery.

5. Ulceration, either of the lining membrane, or of all the coats opening into the pleuræ, pericardium, or communicating with an aneurism of the ascending aorta.

6. Contraction.

Spasm of the Pulmonary Artery and Valves.

In most cases of palpitation connected with disease of the mitral and aortic valves, or the heart, there will be heard, on placing the stethescope over the second or upper part of the third intercostal space, some alteration in the sounds of the pulmonary artery: they can be traced behind the sternum, where they blend with the aortic sounds, and produce murmurs not unlike those met with in aneurism of this vessel. This alteration is constantly met with in nervous palpitation. A careful consideration of the trunk of the pulmonary artery will explain this. The vessel (see the diagram in the chapter on aneurism) is very short; its trunk lies somewhat under and to the left of the aorta, which under certain circumstances may press on it. The right division of the artery, the larger of the two, passes behind the aorta, to form the root of the right lung. In chorea of the heart (see Chapter II.) a blowing sound is particularly liable to occur in the pulmonary artery and valves, whether they are altered or not. This was observed in a case in which there was no permanent alteration, reported at page 11. The palpitation occurred after the first sleep, and at irregular times. There was a distinct blowing sound heard in the course of the pulmonary artery, which was greatly increased in intensity during the attacks of palpitation. He recovered. In a case of a clerk aged 19, reported at page 13, in which there was permanent murmur in the valves, mental excitement increased the sound, opium and morphia had a similar effect. In another case, a lad of 15, growing fast, in which there was a permanent murmur in the pulmonary valves, in addition to the attacks of palpitation, he had difficulty in walking up a hill, or fast, from pain being excited in the second intercostal space, and difficulty of breathing, which compelled him to stop and make two or three deep inspirations. He was then able to go on again. Excessive exertion in rowing a race, and over walking had produced the alteration. This pain and inability to walk far, without taking a deep breath, is often observed in aneurism of the aorta, but it seems generally to depend on pressure on the bronchial tubes. Disease of the mitral valve has a great tendency to produce spasm of the pulmonary valves. As in chorea of the heart, the attacks of palpitation usually occur early in the morning. In an elderly woman who worked hard at washing and mangling, and who became subject to attacks of nervous palpitation, which were relieved by iron and quinine, disease of the mitral valve at length appeared. It is not uncommon in disease of the mitral valve to find alteration in the pulmo-

nary sounds during attacks of palpitation, and even thickening of
these valves. When spasms of the pulmonary valves or of the
artery occur, it is from two causes—(1.) Disease either of the vessel
or its valves ; or, (2.) In connection with disease of the womb or
ovaries—stomach—anæmia—the excessive use of tobacco or
spirits, or chorea of some other part of the body. If disease exists
either in the vessel or its valves, these causes will aggravate the
symptoms. Hence it is not always possible to separate functional
from organic derangement. In the former the spasms are uncertain in their occurrence, the patient in the interval of the attacks
being able to walk fast and even run without difficulty ; but in
the latter there is pain or weight of a more or less constant
character in the left side of the sternum, either confined to the
second or third intercartilaginous space or the junction of the
cartilages with the sternum. With the pain or weight there will
be more less alteration in the valves or in the artery at the point
where it is pressed upon by the aorta behind the sternum. When
in the valves the first sound is either louder than natural or more
or less harsh, the second one being obscured ; but when in the
artery the sound is loudest behind the sternum, and blends with
the systole in a confused manner where the two vessels come in
contact, passing thence up the aorta for a short distance, and
along the right pulmonary artery until it divides. With these
alterations there is more or less difficulty of breathing of an
asthmatic character, but unattended with the mucous or dry
wheezing observed in asthma, or cough, and expectoration ; but
there is violent palpitation of the heart, which like the asthma,
with which it occurs, exists only for a time. There is, as I have
before observed, very great sympathy between the aortic, the

mitral, and the pulmonary valves, particularly between the two last.

CONGESTION AND INFLAMMATION.—These alterations are occasionally met with. In the female aged 22, who died of congestion of the lungs (see page 75) the pulmonary valves were thickened, and the lining membrane of the vessel deeply congested. In another case (see page 79), a male aged 40, suffering from chronic disease of the heart, who died from congestion of the lungs, the pulmonary artery, and the other large vessels were deeply congested. The aortic and mitral valves were thickened, and covered with fibrinous deposits.

In the cases of acute congestion of the lungs which I have had an opportunity of examining after death here, whether existing alone, or in connection with pulmonary apoplexy, or pulmonary gangrene, the first, second, or third divisions of the artery were congested, and the branches leading to the parts in which the apoplexy or gangrene existed were obliterated or closed by fibrinous clots. Some writers speak of the existence of a fibrinous exudation being observed. This is, I think, an exception, the rule being either congestion or fibrinous clots. Dr. Whitley, in a paper published in *Guy's Hospital Reports*, Vol. III., Series iii., reports a case (No. ii.), in which vegetations were found on the pulmonary valves. The pulmonary artery and the right ventricle were distended with clots. The right ventricle was enlarged, and its walls thickened. There were vegetations on the mitral valve; there was a communication between the ventricles. Mr. Paget, in the 27th volume of the *Medico-Chirurgical Transactions*, relates the case of a prostitute aged 20. There were only two pulmonary valves; fibrinous growths adhered to their

H

borders. There were warty growths in the lining membrane of the trunk of the pulmonary artery, and two ulcers, one over the valve, the other at the angle of the division, into right and left artery.

In this climate, inflammation or congestion of the pulmonary artery may occur under three circumstances—either in connection with (1), colonial fever, which never occurs without more or less congestion of the lungs, as the result of some poison circulating in the blood, and in uræmic poisoning; or (2), from either traumatic or idiopathic pleuritis, which may affect the minute divisions of the vessel, and extend to the larger branches; or (3), in chlorosis, or diseases of an anæmic or scorbutic type, the clot forms first in the right ventricle, whence it extends through the pulmonary valves, along the artery; but if unable to do so, it is rolled up into a ball of more or less firmness, or forms an irregular mass, which may be more or less attached or intermixed with the chordæ-tendinæ of the tricuspid valve, and block up the right ventricle. It is very difficult to determine whether the cerebral symptoms observed in colonial fever are the result of blood poisoning, or from the obstructed circulation through the lungs. I have never seen a case recover in which with extensive congestion of both lungs delirium coma has set in. Howel (*London Medical Repository*, 1825),* has published a case in which the pulmonary artery, and the membranes of the brains were inflamed (congested).

---

* There are so few medical books in the colony, that I have not been able to refer to this case. I am indebted to Albers Patholog.: Anatomie, for the notice of this case.

The following is a good illustration of acute congestion of this vessel :—

A healthy looking child, 5 years of age, was taken after bathing in cold water, while the skin was desquamating after scarlet fever, with chills, delirium, and difficulty of breathing. On the 2nd day, when seen there was general congestion of both lungs. Delirium severe; unable to answer questions; heart was acting tumultuously, but there was no alteration in the valvular sounds; pulse 150, irregular; had not passed urine for 48 hours; face congested; hands and feet a little puffy. The breathing gradually became difficult, and it died 72 hours after the attack.

It was with great difficulty that a *post mortem* examination could be obtained. The lungs were deeply congested. The right and left sides of the heart, and the pulmonary artery, as far as it could be traced, and the aorta from its commencement to the point where it passed through the diaphragm, were congested. The congestion in the aorta was most marked in its ascending portion. The substance of the heart, the kidneys, and the liver, were more injected than usual. The bladder was empty.

In adults where life is prolonged, the inflammation when it travels from the heart or affects the vessels in the lungs, produces either pulmonary apoplexy or gangrene, according as to whether the obstruction is more or less extensive. Either of these states may cause the lung to break up, and produce cavity. Inflammation and obstruction of the branches of the pulmonary artery have a great deal more to do with the formation of pulmonary cavity here, than in England, where the cavity usually occurs from the softening of tubercular deposit.

The symptoms of the obstruction are more marked when the
vessels becomes inflamed, as a result of, or in connection with idiopathic or traumatic pleuritis.   The following cases are the best
illustrations I have been able to collect :—

*Spitting of Blood, with pain in the left side of the sternum,
and alteration in the course of the Pulmonary Artery.—
Death.—Clots in the Right Ventricle and Pulmonary
Artery.*

A labourer of florid complexion, aged 30, applied at the institution on March 8th, suffering from slight spitting of blood,
attended with pain to the left of the sternum, near the base of
the heart.   There was some slight tumultuous action heard in the
course of the pulmonary artery ; the heart's action was increased,
but there was no alteration in the valves.   There was fine and
coarse crepitation at the base of the lungs posteriorly.   The
breathing was very difficult, more so than the lung symptoms
accounted for ; the slightest movement aggravated the difficulty of
breathing, and excited violent palpitation of the heart.   The
cough was very slight, and the expectoration scanty, consisting of
a little pink coloured mucus; pulse was 120, feeble; urine and
bowels natural ; tongue rather clean ; skin cool.   He was ordered
to confine himself to bed, and apply a blister to the chest, and to
take some gallic acid, with chloric ether.

March 15th.—I was requested to visit him.   There had been
but little alteration in the symptoms until last night, when he
spat up a large quantity of bloody offensive fluid, very disagreeable to the taste.   Pulse 120 ; mucous râles existed in the right
lung ; the cough was very slight, it was unattended by pain ; the

breathing was still difficult. There was still slight pain in the region of the pulmonary artery, and the sounds heard in it were very confused.

18th.—He had continued to bring up the offensive fluid. The left lung was becoming congested. In the evening the whole of this lung had become affected. In the vicinity of the root of the lungs, both anteriorly and posteriorly, there were coarse râles; the breathing was more difficult than in the morning; nose and lips were cold, and dark coloured; breath cold; mind tranquil; a little expectoration of offensive, thin, dirty mucous. He died in the course of the night.

The right pulmonary artery contained long fibrinous clots, which extended into the ramifications of the vessel. The right ventricle contained a round fibrinous mass of the size of a small walnut. The lining membrane of the right side of the heart, and the pulmonary artery was injected. There were several soft, irregular masses, varying in size from a walnut to a small orange, in the right lung. The larger divisions of the left pulmonary artery were free, but the smaller ones were obstructed over a larger space, in the upper and central part of the superior lobe; the lung tissue was soft, and infiltrated with dirty offensive fluid. This case is a good illustration of the fatal form of acute inflammation of the pulmonary artery sometimes met with in the colony, and which invariably ends fatally, when the vessels of both lungs are affected. The extreme difficulty of breathing, which the state of the lungs does not satisfactorily account for, the slight cough, and the scanty expectoration, are very indicative of the nature of the disease. In this case, the round fibrinous clot in the right ventricle would possibly account for the alteration heard in the course of the pulmonary artery.

The next case presents some peculiar features, from an hydatid cyst opening into the bronchial tube.

The patient, 42 years of age, a messenger in the Legislative Council, was supposed to have either received a blow, or some injury of his side, from being jumped upon when on the ground last Boxing Day. He began to complain of pain in his right side, a few days afterward, immediately under the ribs.

On the 4th January he was suddenly seized while working with severe pain in this side. It was relieved by a sedative draught. On the 6th the pain returned. It was again relieved. He had several similar attacks, but was relieved by the treatment adopted. He was, however, very careless, and exposed himself to cold.

Jan. 16th.—He presented the following state :—Pain of a cutting nature in the right side, with marked dulness on percussion, extending posteriorly from the lower angle of the scapula to the first floating rib, and anteriorly from the nipple to one finger's breadth below the false ribs. There was no increased vocal resonance, but fine crepitation existed all around and above this dulness, for a space of from two to three fingers' breadth; posteriorly there was slight ægophony. The left lung was free; pulse 90; urine loaded with lithiates; motions natural. His respiration was more difficult than could be accounted for by the symptoms. There was no alteration about the heart. His face was calm, and his spirits good. He had had slight jaundice, but this had been removed by calomel, and blistering the side. The pain, and the alteration in the respiratory sounds had, Dr. Gregory observed, extended upwards. A blister ordered to the chest, with bromide, and iodide of potash, squills, and decoction of senegæ. His mouth was slightly affected, from the calomel which Dr. Gregory had so judiciously given him.

25th.—Pulse same ; respiration the same, although the crepitation seemed less ; cough very slight, and but very little expectoration of a little clear mucus ; skin cool ; urine natural. In the posterior part of the right lung, there was a coarse râle heard in the course of a large bronchial tube. The respiratory sounds in the left lung were rather feeble. The heart sounds were natural. There was a little coarse mucous râle in the left lung, in front, in the vicinity of the base of the heart, loud enough to mask any sound in the pulmonary artery. His face was calm; spirits good The only unfavourable symptom was the difficult respiration.

26th.—In the afternoon, he brought up from the lungs a large quantity—about a pint—of thick glairy, offensive, and disagreeable fluid, like that found in an old hydatid cyst. Dr. Gregory saw him soon after; he was very much collapsed. By the exhibition of stimulants, and the application of bottles filled with hot water, the circulation and warmth of the skin were restored. In the evening, five hours after the occurence of this attack, his face was calm ; skin cool ; pulse 130, feeble ; the respiration was very difficult. Coarse mucous rales existed in both lungs, intermixed in the left one with fine crepitation. In the posterior part of the right lung, between the spine of the scapula, and its angle, there was an obscure, cavernous sound. He was expectorating a little offensive, disagreeable white mucus, tinged with blood. Blistering fluid was applied to the front of the chest. On the morning of the 27th he seemed much better, the blister had given him considerable relief. The expectoration the same, very scanty; pulse 120. In the evening there was but little alteration in the symptoms.

28th.—He had passed a restless night. The lungs were more gorged; no expectoration; pulse had again risen to 130; turpentine was now given! it caused copious expectoration. He died rather suddenly, at 4 p.m.

His friends would not allow the body to be examined.

In the next case there was no history during life to be obtained.

He was a labourer, about 40 years of age, but of late had lived on the earnings of his wife. He had been to the hospital for something to relieve his breathing, which had been difficult on exertion for more than a week before his death. For three days before his death he had kept his bed. He had no cough or expectoration, and only complained of oppression in the chest. Not being generally inclined to work, he was thought not to be ill. While attempting to dress himself he fell backwards, his face became death-like, breathing short and hurried, and died in a few minutes. He was seen by my late friend, Dr. Stewart, immediately after death, who kindly requested me to help him to make the *post mortem* examination. The brain and membranes were deeply congested. The lungs were slightly adherent, and each pleural cavity contained from 8 to 10 ounces of fluid. The pericardium contained a little more fluid than usual. The heart was flat and flaccid. The left ventricle was empty, its walls soft, and rather fatty; the right one contained soft, fibrinous clots, sufficient to fill a gill measure. A clot was easily withdrawn from the pulmonary artery, to the length of several inches. On tracing the course of the pulmonary artery, in both lungs these clots were found. They were more intermixed with the colouring matter of the blood in the smaller divisions of the vessel than in the larger. The liver was in a state of fatty degeneration ; the kidneys the same. There was an old hydatid cyst situated between the right kidney and the liver. It was of the size of a large orange.

In the two following cases, the inflammation was traumatic depending on fracture of the ribs, in one recent, in the other of long standing with caries.

Fracture of the Ribs—Death on the third day, after a hearty meal—Congestion of Vessels of Brain—Heart empty— Clots in the Pulmonary Artery.

A man between 40 and 50 was brought into the hospital, with fracture of five or six of the left ribs, from being run over by a water cart. He died suddenly on the third day, after eating a hearty breakfast. The neck was very much congested. The vessels of the scalp and brain were congested ; there were a great many blood points in the substance of the brain ; the lateral ventricles, particularly the left one, contained serum. The arachnoid membrane adhered very closely to the dura mater, on both sides of the fissure of the cerebrum. The heart was large and flabby ; its cavities were empty. The pulmonary arteries contained long fibrinous clots. The stomach was very much distended with food and flatus, which seemed to have pressed up the heart. He had had dysentry for some time, and had only left the hospital a short time before he met with the accident. The lower part of the colon, including the sigmoid flexure for the length of 30 or 40 inches, was thickened, and large veins ramified on its surface. He made no complaint beyond saying that the bandage felt tight, although it was far from being so. His breathing was difficult. It was thought that his lung had been injured, but he spat up no blood, and had no cough or expectoration.

Caries of the ribs—Death—Clots in the Pulmonary Artery and Left Auricle—Disease of the Aortic and Mitral Valves— Liver and Kidneys diseased.

A middle-aged man was brought into the hospital suffering from caries of the ribs, the result of an old fracture. The end of one of the ribs had penetrated the left side of the mediastinum. The pus from the carious ribs had penetrated along the intercostal muscles, and opened near their angles. The pulmonary arteries contained long clots of very dense fibrine near the heart, which become darker as it receded, and softer, as if deprived of some of its fibrine. The left auricle was distended with a very large clot of blood. The other cavities were empty. The tendons of the mitral valve were adherent two and three together, and dense and cartilaginous. The fleshy columns were not lengthened. The aortic valves were thickened. The aorta as high as the arch was studded with ætheromatous deposits. There were also deposits on the internal surface of the pulmonary artery. There was a softened space in the apex of the right lung, of the size of an apple; the rest of this lobe was indurated. In the apex of the left lung there were indurated masses with spots of pus in their centres. The posterior parts of both lungs were congested; the tissue of the right one was softened. The lung on the left side adhered to the ribs. The liver was dense and hard, but not markedly cirrhosed. The kidneys were both enlarged, and covered with a network of vessels. There

was a large quantity of granular matter in their tubular
substance.

This patient was brought into the hospital by the police.
No account could be obtained of his history previous to
admission.

Dr. Kidd, in a very valuable paper in the *Dublin Journal of
Medicine*,* thinks that the formation of the clots in the pulmonary
artery is attended by a series of symptoms so well marked as to
draw attention to the condition of the vessel. He says that there
is first a degree of fever with impaired appetite, scanty, or suppressed secretions, and vomiting, such as usually occur at the
onset of phlebitis elsewhere; a sense of pain at the pericardium,
and tightness of the chest; the face assumes the well-known
phlebitic aspect; great debility, attended with a tendency to
fainting; the pulse quick, weak, and soft. The debility is greater
than the other symptoms account for. The languid circulation,
and the congested state of the nervous system, are still further
shown by the violet patches, the œdema, and tendency to slough-
ing of the integuments. The over-burdened right ventricle
relieves itself by regurgitation, and hence, perhaps, the swollen
and pulsatory jugulars. The respiratory movements are rapid,
attempting to compensate by increased action for the limited
portion of the lung that is taking part in the aëration of the blood.
The constitutional symptoms may now increase in intensity, mil-
liary or red eruptions appear, and the patient gradually falls into
the typhoid state and dies. All this time the uterus, if it be a
puerperal case, may show no signs that its veins are affected. As
in Curveilhier's case, the disease may have been relieved by early

* Rankin's *Retrospect of Medicine*, Vol. XXIV.

treatment. In 32 cases which occurred from various causes, death occurred suddenly, that is, either in apparent health, or in apparently favourable convalesence, in 10 instances. In 6 it occurred during serious illness, but there were no symptoms to indicate the existence of clots. These observations of the doctor's do not apply to all cases. The fever may be so slight as to escape notice, and when it exists may depend on inflammation of some other organ or part of the body, as the uterus iliac, or femoral veins. A sense of pain, not in the pericardial region, but at the base of the heart, is characteristic, and must always be looked upon with suspicion. The difficulty of breathing depends upon the extent with which the circulation is interfered with through the pulmonary artery, and is not to be accounted for by the symptoms referable to the lungs; the presence of a murmur or a churning sound must be looked upon as strong evidence of the formation of clots either in the right ventricle or in the pulmonary artery. Obstruction of the pulmonary artery or of the right ventricle from fibrinous clots is more common here than it is supposed to be. The symptoms preceding the fatal syncope are often so slight here that they escape notice. The following brief notes of cases which have proved fatal from the presence of a clot in the right ventricle, may not be uninteresting:—

I. A chlorotic servant girl (under 20 years of age), at a baker's in Church-street, Richmond, died suddenly. The brain was pale and its vessels empty. The right ventricle contained about two ounces of soft fibrinous clot of a very pale colour. There was scarcely any blood in her body. The left side of the heart was empty. There was no food in the stomach. The stomach adhered to the pancreas around a small ulcer, which was of the size of a sixpence, with regular edges. The edges of the ulcer were quite smooth.

II. A female between 30 and 40, the mother of several children, was confined by a midwife, who applied cold wet cloths to the abdomen, although there was no hæmorrhage. On the third day she presented the following symptoms:—Pulse feeble (120) discharge from the uterus rather scanty. There was no pain or tardiness of the abdomen, but she had had several attacks of faint-ing. There was no alteration in the left side of the heart, but there was a feeling of uneasiness near its base, and to the left of the nipple. At this point there was a distinct churning sound. She died suddenly on the fifth day. She had been for some years subject to palpitation of the heart. A brother had died from disease of the heart. The lungs were pale and rather contracted, but otherwise healthy. The left side of the heart was empty. Its walls and the walls of the right side were thin. The valves were healthy. The right ventricle contained a large soft clot. The pulmonary arteries were free. The lining membrane of the arteries presented no alteration. The cold wet bandage seemed to have depressed the heart's action so much as to favour the formation of the clot. It is common to see here patients sink under the formation of clot in the pulmonary artery, and in the right ventricle, if the vital powers have been very much depressed, either by repeated attacks of flooding, or hæmorrhage, in diarrhœa in drunkards, or in those whose constitu-tional powers have been very much reduced. The pulse is very feeble and quick. An examination of the right side of the heart will show either a churning sound from the clot being involved in the valves, or there may be, as when a clot forms in the aorta, a slight musical sound. I have repeatedly watched the formation of the clot in the pulmonary artery, and found that it takes about 72 hours in the aged and debilitated to form. It would seem that a want of power in the right ventricle or in the pulmonary artery, or some alteration in the blood itself, will favour its formation. In a paper published in 1861, in the *Medical Journal*, I drew attention to the occurrence of fibrinous clots in the aorta from the presence of urea in the blood.

It cannot, I think, be considered that the occurrence of fibrinous clots is altogether due to inflammation. When fibrinous clots form in the debilitated, the urine is often very scanty. It may possibly depend on the blood undergoing some alteration which renders it liable to coagulate.

There are several highly interesting cases recorded in which the artery became obstructed by cancerous masses, one by Dr. Wernher, the others by Mr. Paget. In the former gentleman's case, the patient, a man aged 22 years,* suffering from encephaloid cancer of the knee, and tibia, was taken on the 27th of January with pain in the region of the heart, to the left of the sternum. Difficult and hurried breathing. There were no signs the next day of alteration in the heart, and no cough or expectoration; pulse 140. In the night he took another attack of difficulty of breathing, his pulse could scarcely be felt. Third day he spat up some coagulated blood. He then got better, and was as usual. The leg was removed on the 7th February. The next day there was considerable fever. On the 10th there was a return of the spitting of blood, which increased up to the 19th. Then there was dulness

* Henle's Zeitschrift, 1855, cited in Vol. 21 of Rankin's Abstract.

on percussion at the base of the lungs, with shivering. 20th—Cough and expectoration of bloody offensive serum and mucous râles all over the chest, with shiverings, fever and milliaria. 24th—Taken with symptoms of asphyxia, became collapsed, and died. The branches of the right pulmonary artery contained coagula of coherent cancerous masses. The left pulmonary artery and the pulmonary veins were free. The lungs contained large gangrenous masses. In Mr. Paget's case,* the patient, a female aged 45, a drunkard, died from disease of the liver, with dropsy of the legs and abdomen. The liver was enlarged, hard and yellow. It contained several soft medullary tumours ; the tissue of the organ and the hepatic vessels, the same, but it was more fluid, and of a bright yellow hue. The substance of the lungs contained cancerous masses unconnected with its substance. The pulmonary artery contained a similar yellow substance. When a section of the lung was made, fine yellow lines were seen, as if the smaller branches of the artery had been injected with chromate of lead.

In two other cases† similar deposits existed in the vessel, but it was uncoloured.

In another case ‡ the trunk of the left pulmonary artery was filled with a large firm mass composed of fibrine mixed with cancerous masses. The left lung was adherent and congested; the right one pale and ædematous. There were no cancerous deposits in the lungs. Mr. Lawrence in eleven years had removed both her breasts, and three times growths from the cicatrices. She died suddenly while apparently doing well.

* Medico-Chir. Trans.: Vol. 27, 165.
† Disease of Lungs, No 19.—St. Bartholomew's Museum.
‡ Ibid, No. 100.—Ibid.

Cancer is much more common in the colony than in England. It is oftener encephaloid than squirrhoid; its growth is therefore more rapid, and secondary deposits more liable to occur. Deposits in the pulmonary artery, when the cancer is diffused and soft, may be expected to be met with from their being arrested in one or other of the pulmonary arteries according to the size of the mass. The same circumstance may occur in phlebitis, when portions of a clot are separated and become arrested. If a large sized vessel is obstructed, gangrenous abscesses may be excited in the lung, or the clots may form a nucleus for the deposit of fibrine, which will be more or less dense according to the time they have existed before death occurs. If the red corpuscles of the blood cannot escape, then the obstructed vessel will be found to contain a redish defibrinated fluid, distinct from the fibrinous clot. These alterations resemble very much the changes found when blood is "whipt," the fibrine more or less dense separates, leaving the colouring matter and the serum.

The two following cases, in which the clot seemed to be conveyed into the pulmonary artery, may help to explain the circumstances under which clots sometimes form in the pulmonary arteries:—

The wife of a shoemaker, aged 24, the mother of several children, was confined on the 13th of May, in a small, overcrowded house. She did well until the eleventh day, when under some mental excitement, and while making her bed, she was taken with severe flooding. It was checked with considerable difficulty, by pressure. At the end of seven days, she was so far recovered as to be able to be removed to another house; suddenly, without any notice, the right leg presented symptoms of phlebitis.

By leeching, the constant application of hot bran poultices, with iodide of potash internally, the swelling of the leg subsided rapidly, but the femoral vein remained hard, swollen, and painful. The womb felt enlarged; it was free from pain or tenderness; there was a little offensive discharge from it. The constitutional symptoms were very slight. In the evening of the ninth day, she complained suddenly of pain in the left side of the sternum, with palpitation of the heart. She became hysterical. In the morning the heat of the skin was increased; the pulse had risen to 115; it was very feeble; tongue white; she had passed no urine; the respiration was unusually quick. She had had no discharge from the womb for three or four days. The weather being cold and wet, it was thought that she had got a chill. The pain was referred to the base of the heart, immediately above, and rather to the right of the nipple; here there was a confused sound, which could be traced along the pulmonary artery. This could be easily done from there being no alteration in the mitral or aortic valves. Ten drops of turpentine were given on a lump of sugar, every two hours, and the chest painted with blistering fluid. She was ordered to be kept perfectly quiet, and not to sit up. In the evening she was suffering considerably from stranguary from the blistering fluid. She had attempted to sit up, but had fainted. On the morning of the third day, she was much better, the respiration was not so quick; the pulse had fallen to 110; it was fuller. The turpentine produced copious perspiration, and increased the quantity of urine. She had so far recovered at the end of a week as to be able to sit up without difficulty. Her convalesence was slow. My attention was particularly directed to this case from one which came under my care in London.

I

The patient, the wife of a cabman residing over a confined, offensive stable, was taken with flooding, from having intimacy with her husband ' on the ninth day after her confinement. She got subsequently swelling of one of the legs, with offensive discharge from the uterus. She had fever; dry, brown tongue; slight delirium, with pain in the chest. She was attended by one of the pupils of the Dispensary, who had confined her. There seemed no reason for anxiety. On the 21st day, she was much better than usual, got up, dressed herself, and while attempting to move the cradle, fell down, and expired in a few minutes. The longitudinal sinus contained a long, thin clot; the vessels of the brain were rather injected; the membranes were a little opaque; the brain itself was healthy. The pulmonary artery throughout contained a rather firm clot; the clot was firmer the further it was from the pulmonary valves. There were distinct evidences of small, fibrinous granules in the smaller divisions of the vessels, and it appeared as if they had been the nucleii on which the fibrinous clots had formed. The pulmonary artery was very much contracted, its lining membrane was dark coloured. The right ventricle was contracted; the left auricle and ventricle contained dark clots. The uterus was not enlarged, but it was soft, and from its cut surface a thin, reddish coloured serum exuded. The left illiac and femoral veins contained clots. This leg was œdematous.

3. *Dilatation of the pulmonary artery* :—

This is an alteration very likely to be met with in the colony. The dilatation may be either general, affecting the trunk and sometimes the right or left branch, sometimes both, or the main artery, which may be saculated. The general cause of dilatation is pressure of an enlarged aorta, aneurism, or cysts, or abscesses in the lungs. It is possible that chronic pulmonary emphysema, and 'even hooping cough may cause dilatation. Any obstruction in the course of the vessel would necessarily call upon the right ventricle to use more force, with the result of producing more or less hypertrophy and dilatation of its cavity. The early writers seem to have been aware of this alteration. Ambroise Paré* found *l'artère veineuse* so much dilated that the fist could be introduced; its internal surface was ossified. In the Ephemerides,† it is said, "*Arteria pulmonalis tam copioso sanguinie turgescebat, ut, quasi anevrysmate affecta, præter propriam magnitudinem præternaturalem, hinc inde succulos ernore coagulato turgidos habuerit appensos.*" Morgagni‡ found this vessel and the aorta dilated in a woman 64 years of age, who had suffered from difficulty of breathing, with dropsy of the upper and lower extremities. The lungs were congested, the pleural cavities and the pericardium contained yellow serum. In a man who died with symptoms of disease of the chest, he found the pulmonary artery and aorta dilated. The inner surface of the aorta was hard and irregular. *Epist.* lxiv. In another case, the patient, 59 years of age, was subject to attacks of fainting. He died suddenly. The aorta had given way, and blood had escaped into the pericardium. The pulmonary artery and the aorta were enlarged.

Bonet,‖ in a woman who had suffered from palpitation, says— "*Arteria venosa, quia in monstruosum extensa erat, humore seroso*

---

* Laennec Traite de l'Auscultation, II., 730.
† Dec. iii. an. v., obs. 207, cited by Laennec.
‡ *De Sed: et Caus. Mort.*, xxiii.—6.
‖ Sepulchretum, tom I., obs. 27.

impleta." Lacnnec* considers that this vessel is rarely dilated. He often noticed it enlarged in old cases of disease of the chest. In one case he could introduce three fingers into it. Hope† found the vessel four and a half inches in circumference. Albers‡ alludes to a case in which it was enlarged.

The following are the best of the cases I have been able to collect :—

Symptoms of Dilatation of the Pulmonary Artery—Relieved by Treatment—Death.

A labourer, 43 years of age, applied at the institution on the 12th September, suffering from pain in the left side of the sternum, near the cartilage of the third rib, passing through to the back. Immediately over the seat of the pain there was a loud murmur, which extended upwards for a short space, when it was lost. He had great difficulty in walking, being compelled to stop and take a deep inspiration. He was then able to go on for a short distance, when he was again obliged to stop. He could not lie down low on the bed on his back, and he had some difficulty in lying on either side for any length of time. The pulse was equal in both wrists—90, but feeble. The heart's sounds were natural ; there was no alteration either in the aorta or in the carotids. The external jugular veins were not markedly enlarged. The aorta could be just felt behind the sternum. His face was of a waxy hue. He had drunk and worked hard, but never had rheumatism. His liver extended from one finger's breadth below the nipple, to the same distance

---

* Tome II., 729.

† Diseases of Heart, 420.

‡ Anatomie Pathologique.

below the false ribs. Digitalis, with bromide and iodide of potash, and ammonia were given. Strict rest ordered.

Oct. 1st.—He was much better; the pain between the shoulders and on the left side of the sternum had disappeared; he could walk better, and lie longer on his back. The bruit had diminished very much in intensity. He ceased to attend. He died suddenly, about three months after this. I could not learn what were the *post mortem* appearances.

In the cases reported at page 87, there was dilatation both of the aorta and pulmonary artery. If the external jugular veins enlarge, it is generally from pressure on the descending vena cava, by the aorta. This enlargement is not to be considered as symptomatic of dilatation of the pulmonary artery alone. I have seen these veins in aneurism of the right side of the aorta, just below the arch, as large as the little finger.

The next case presents points of great interest:—

Supposed Asthma occurring after Fever, with Congestion of the Lungs—Symptoms of Dilatation of the Pulmonary Artery—Death.

A farmer, working hard, 45 years of age, applied at the institution, on the 21st November, suffering from "asthma." The attacks did not occur, as they generally do here, towards morning, but on exertion, and particularly after a heavy meal. The respiratory sound was rather feeble in both lungs, but it was more marked in the right than in the left lung; but there were no other signs referable to the lungs. He could not walk fast, and particularly up a hill, without being compelled to stop and take a deep breath. The heart's sounds were natural, and there

was no increased impulse. The symptoms seemed to indicate
that there was an aneurism of the thoracic aorta. His legs
were slightly swollen, but there was no albumen in the urine.
The liver was rather large. He had not drank for the last six
months. He thought that the disease was first excited by sleeping
in his cart in the cold and wet in June. He had then fever, with
cough, difficulty of breathing, and expectoration tinged with
dark-coloured blood, and mixed with clots of blood of a dark
colour. He got gradually better. But as soon as he began to
get about and work, the difficulty of breathing set in. The
heart's dulness at its base was increased in extent ; the organ was
not displaced, as it often is when the liver is enlarged.

There was a slight murmur on the mitral valve with the first
sound. The aortic sounds were natural, but on tracing the aorta
up along the right side of the sternum close to the upper border of
the cartilage of the third rib, there was a kind of confused whiss-
like sound similar to that sometimes heard when the aorta is bent
on itself. The sound was not carried up along the aorta beyond
an inch and a half, nor could it be heard in the vessel below for
more than an inch. It could be followed along the upper border of
the right cartilage of the rib, but it was soon lost; on the left side of
the sternum the sound was feeble, increasing in intensity towards
the right side. On making him walk quickly up and down the
room, until what he called the " panting for breath " was produced,
the sound was increased in loudness, the pulse rose from 86 to
120, but it was feeble. He then complained of the sense
of deep constriction, which he always had, becoming painful.
The veins of the neck during this state were a little enlarged; his
head felt full, and his face flushed. He had a strong tendency to

vomit during this state, and for some weeks, as soon as he got out of bed in the morning, he vomited whatever food he had taken the previous night, or if his stomach was empty, "slimy stuff." He could not lie on his back with his head low; if he did, the pain and constriction in the chest would set in, with fulness of the head and retching. He could rest best on his left side, with his chest not very high and bent a little forward. If he eat a very hearty meal before going to bed, he could not rest, for on changing his position an attack would be excited. To get relief, he was obliged to excite vomiting. The vomiting had reduced him very much in strength. His face was pale and sallow. By strict rest, beef tea, injections instead of giving food by the mouth, and small doses of iron and digitalis, he so far improved in the course of five weeks as to be able to walk a short distance without difficulty. The sound to the right of the sternum, and the sense of constriction, had very much diminished. He returned home. There he walked about, drank rather freely of spirits, not being able to eat. He was brought back again to town on the 27th of December, with all the symptoms aggravated. He was unable to retain any food on his stomach. There was a strong tendency to diarrhœa, which prevented the beef-tea injections being retained. His legs were very much swollen, and his abdomen contained fluid. He sank gradually, and died at the end of 10 days.

The sternum, ribs, and the half of the clavicles were removed, to get as good a view as possible of the heart and large vessels. The aorta, below the arch, was larger than at the arch. The trunk of the pulmonary artery, and its right branch, were very much dilated. The former formed a kind of pouch, large enough to receive a small orange. The coats were very thin. The pulmonary valves were unusually long. The heart was flabby. The walls of the right ventricle were very thin; its cavity was unusually large, and seemed to form quite two-thirds of the organ. These were the only changes worth noting.

Dr. Whitley has reported a case in which there was *fremissement cataire*, and a murmur in the veins of the neck and over the clavicle. The case is highly interesting :—

A female aged 44, entered March 11th. She had had, 13 years before, an attack of paralysis; her left hand was still weak. Ten months before admission, when pregnant, she fell backwards, and for fourteen days spat blood, and passed bloody urine. She then became dropsical, until she was confined prematurely, when it disappeared, but returned at the end of two months with cough, difficulty of breathing, and increasing emaciation. On admission, she was anæmic; face anxious. She complained of pain in the left side of the sternum, between the second and third ribs. There was well-marked *fremissement cataire*, over a space extending from one to three and a-half inches below the left clavicle, and three inches from the left border of the sternum; the point where it was most marked was three-fourths of an inch to the left of the sternum, and one and a-half inch below the clavicle. There was a loud, rushing sound between the second and third cartilages on the right side of the sternum, with the first sound; the second sound was natural. The veins in the neck were congested. There was a soft murmur in the veins of the neck, and over the clavicle; pulse 84, soft and compressible; urine albuminous; abdomen contained a considerable quantity of fluid. On the 17th, she was unable to lie down, from danger of suffocation; vomiting set in, and on the 20th, death. The pulmonary artery and its right and left branches were very much dilated. The lining membrane was covered with ætheromatous deposits. The right ventricle and the pulmonary artery contained a dark-coloured, semi-fluid clot. The lungs were congested, approaching almost to pulmonary apoplexy. The right ventricle was dilated. The superior vena cava, and the jugular veins were distended with blood. One of the folds of the mitral valve was thickened. There was no other change. It is therefore to be presumed that the aorta was not dilated, as was the case in the patients observed here. Dilatation of the aorta is not uncommon here.

4. *Deposits and thickening.*—Morgangni, in a man aged 60, who died with symptoms of disease of the chest, found the pulmonary valves thickened. The mitral and aortic valves, the same; the aorta was dilated, and its coats thickened. The left ventricle was dilated, but its walls were not thickened. Valsalva found the pulmonary valves so united as to leave only a small opening of the size of a lentil, from the obstruction to the passage of the blood; the right cavities of the heart were very much dilated. The mitral valve was incomplete, and the foramen ovale open. The patient, a female, was 16 years of age.* Meckel, Albers states, met with a similar case in a girl 14 years of age; the foramen ovale had remained open, the pulmonary artery was dilated. Portal found the pulmonary artery and its valves thickened, and Chomel, the artery ossified. From deposits in the aorta being more frequently met with here than in Europe, and from colonial fever being attended with more or less congestion of the pulmonary vessels, they may be expected to be found. Sandifort, Stoll, Lobstein, Bichat, Andral, Whitley, and others have found deposits in this vessel. Laennec says—"*Il*

* Morgangni, Epist. xxiii.

existe a piene trois à quatre exemples d'incrustations osseuse dans l'artere." Hope, as before observed, found but few cases. In a case observed by myself, there were deposits in the pulmonary artery, and in the aorta. A case occurred to myself, in which there was a dense cartilaginous ring around the pulmonary artery, immediately at the point where it sprang from the right ventricle. The pulmonary valves were involved in the alteration; the heart was very large and flabby; the walls of the ventricles were not thickened, but their cavities were very much dilated, particularly the right one. This ventricle and the pulmonary artery formed one cavity, the opening readily admitting the four fingers brought together, as high as the second joint of the little finger. The pulmonary artery was dilated above the cartilaginous ring. The patient was about 50 years of age, a printer, working in the office of Messrs. Mason and Firth. He attributed the occurence of the disease to a blow on the chest. For some years, he had a constant dull deep-seated pain just above the nipple, becoming cutting on the slightest exertion, and attended with great difficulty of breathing, with a tendency to faint. These had gradually increased in severity. His pulse was always feeble—exertion rendering it still more so. The cardiac dulness was unusually extensive, but there was no alteration in the sounds of the heart—they were very feeble. He sank gradually, worn out by the pain and loss of sleep, from not being able to lie down.

5. *Ulceration of the Pulmonary Artery.*—This is of rare occurrence. It may be met with—

First—As simple ulceration of the lining membrane of the vessel, without the escape of blood.

Second—Ulceration, with transudation of blood, or rupture of the coats of the vessel; the blood escaping into the pericardium, the pleural cavities, cavities near the root of the lungs or into cysts.

Third—From the pressure of an aneurism of the ascending portion of the aorta. The walls of the aneurism, and the coats of the pulmonary artery unite, and either from ulceration or extreme thinness, the coats may give way, and a communication form between the two.

1. *Simple ulceration.* This is very rarely observed. Mr. Paget found two in the lining membrane, one half-an-inch in diameter, behind one of the pulmonary valves; the other at the angle of the bifurcation, but smaller in size. There was thickening of the valves, with soft yellow fibrinous exudation attached to their free borders. There were only two valves. The patient, a prostitute, was 20 years of age. In the next case, the blood exuded through the ulcer. The patient, aged 32, a carpenter, of very sober habits, was taken, while at work, after his tea, with pain in the region of the heart. When seen by my friend Dr. Daniel (who very kindly brought the case under my notice), he was nearly pulseless and skin cold. The hearts' sounds he found were unusually obscure. He died in an hour from the time he was taken ill. His wife informed us that a month before, while loading a cart with dirt (work he was not accustomed to), a sudden pain occurred in the region of the heart; it lasted for some time. We found the heart surrounded by a black clot, an inch in thickness. Immediately above the pulmonary valves, there was a deep red spot the size of a sixpence. Externally it was covered by a thin transparent exudation; below this there was a vascular space distinctly defined. The internal membrane was wanting. This vascular space was soft and easily

broken up. The blood had exuded through it. There was no other alteration.

Usually the vessel gives way, although even this is of rare occurrence. Tabarrno* found in the body of Cardinal Boncompagno, the pericardium distended with blood from an opening in the pulmonary artery. A soldier† about 20 years of age, after complaining for about 3 months with slight difficulty of breathing on exertion, was taken suddenly with severe pain soon after going to bed, on the left side of the sternum, with great difficulty of breathing, coldness and faintness, the body being bathed in perspiration. He rallied, but complained of feeling weak, with pain in the region of the heart, and palpitation. He was bled with considerable relief.

At 11 a.m. he complained of exhaustion; his face got pale—pulse ceased, and he died in a few minutes.

The pericardium contained clots of blood. The internal surface of the pulmonary artery was covered with a layer of fibrine. Just above the valves there was an opening, and it was through it that the blood had escaped into the pericardium. The coats of the vessel in the vicinity of the opening were deeply congested. The lungs were pale; the heart was healthy, except the left ventricle, the walls of which were rather thicker than usual.

The artery opening into a cavity of the lung—this is of more frequent occurrence than is supposed, particularly when there is a cavity near the root of the lung. Mr. Crowfoot has reported an unusual case in the 26th volume of the Medico-Chirurgical Transactions, in which the artery ulcerated and opened into a cavity in the left lung. Mr. Beales says that the vessel was as large as the

* Morgangni.
† Cospar's Wochenschrift, 1842.

aorta; it curved upwards as high as the clavicle. The opening, which was as large as a crow's quill, was two inches from the bifurcation.

The formation of an opening between the aneurism of the aorta and the pulmonary artery, is more frequently met with than ulceration and perforation. From aneurism of the first part of the aorta immediately above the valves being common in the colony, it must be expected to be met with oftener than in England. Death may either take place suddenly, or the patient may live for some time, the admixture of venous and arterial blood may cause symptoms, more or less marked, of cyanosis. When the patient lives for some time, the opening becomes more or less rounded off and cartilaginous; but when it takes place suddenly, it is rugged. I have seen one case in the hospital in which the aneurism, a small one springing from the deep aspect of the aorta immediately above the valves. It adhered closely to the pulmonary artery, the coats of which were very thin; but it opened into the pericardium. I am acquainted with another case in which there was an irregular opening between the aneurism and the artery. The patient, a respectable man, 45 years of age, of irregular habits, had received a blow when drunk in a quarrel in a house of ill fame. He made no complaint at the time. He was taken home in a stupified state. At the end of the twelve hours he became sensible, and then complained of pain in his chest. He died suddenly at the end of forty-eight hours. In the case of a man aged 53,* who died suddenly after suffering for some months with bronchitis and præcardial uneasiness, was suddenly taken with violent dyspoma, followed by insensibility and death in four

* Thurnam.

minutes. There was an aneurism of the ascending aorta, which opened into the pulmonary artery by an irregular jagged opening.

In another case,* the patient aged 53, who had had when 30, symptoms of phthisis—at 39, an attack of paralysis of one side, which left the foot of that side cold. When 50, he began to suffer from flatulence, noise in the ears, with pains and sometimes slight swelling of the hands and feet. While playing with a child between 8 and 9 p.m., he was taken with weight in the chest, and soon after vomiting. He went to bed. About midnight he complained of feeling cold; coughed up some mucus tinged with blood; cold sweats broke out, and his pulse became feeble. When seen at 5 a.m. by Dr. Baillie, his pulse was full and irregular, breathing difficult, face pale, covered with cold sweats and he tossed about as if in great pain. He died soon after.

An aneurism the size of an orange was found at the commencement of the aorta; it adhered to the pulmonary artery just before it divided; here there was a narrow opening half-an-inch in length between the aneurism and the vessel.

In another case, the patient, aged 36,† was taken six months before his admission into the Royal Infirmary under Dr. Alison, with difficulty of breathing and palpitation of the heart. At the end of four months, dropsy set in. His pulse was small and weak. There was a double murmur with the diastole. Fourteen days after admission, he was suddenly taken with faintness. All the symptoms increased in severity. By the exhibition of drastics, the dropsy was kept under. Five weeks after admission, his lips, before pale, became livid, and his pulse feeble. He died in the

* Ibid.   † Ibid.

night. There was a large quantity of blood in the pleural cavity. The heart was enlarged; two of the aortic valves were so elongated that they could not act. There was an aneurism of the size of an orange above the aortic valves. It communicated with the pulmonary artery one inch and a-half above its valves, by a ragged fissure an inch in length.

In the few cases recorded in which death did not take place immediately, the extent of the lividity depended on the size of the opening between the two vessels. In one case communicated by Dr. David Monro,* of Edinburgh, to Dr. Hope, the aorta was dilated from its origin to its arch into an irregular sac, which adhered closely to the pulmonary artery and communicated with it by two openings with round cartilaginous edges, situated an inch and a-half above the valves; one would admit the point of the little finger, the other a crow-quill. A third opening, with ragged edges, existed near the arch. The man, a porter, aged 24, entered the Edinburgh Infirmary, October 30th, 1833, suffering with great difficulty of breathing; face somewhat *livid*, and general dropsy. His pulse was large, harsh and thrilling, 112; heart's action tumultuous and diffused over a large space, with much dullness on percussion. "The first sound was accompanied by a loud soufflet, audible over the whole of the fore part of the chest, particularly at the middle of the sternum and in the back on both sides of the spine. The second sound was short and much obscured by the first; a continuous murmur extending from the first over the second sound." He was temporarily relieved by a slight bleeding, but not by medicines. The pulse became intermitting some days before death, which happened a fortnight after

* Page 469.

admission. Dr. J. H. Bennett has published a case* in which the patient lived for some time after the occurrence of the communication. The patient, aged 33, a teacher, had enjoyed good health until February, when he felt a pain like the pricking of a pin under the ensiform cartilage, which continued for a week. Fourteen days later, he felt something give way in this part while walking; it made him feel weak; but at the end of a fortnight, he was able to follow his occupation. During the next two months, he spat blood occasionally, and suffered from palpitation of the heart. In April, his feet began to swell and his abdomen to enlarge. The palpitation increased with considerable dyspnœa, cough and occasional vomiting. There was a soft murmur at the apex of the heart with the first sound; the second sound was natural. At the junction of the third costal cartilage with the sternum, the first sound was loud, prolonged, and blowing; the second, short and rasping. Over the manubrium of the sternum, there was a rough continuous blowing murmur occupying the period of both sounds; it was heard, but feebler under both clavicles; pulse 90, irregular. He died on the 23rd of June, twelve days after admission.

The skin and eyes had a yellow tinge; general dropsy existed, and phlyctenæ filled with bloody serum on the body and thighs. There was an aneurism of the size of a walnut in the aorta immediately above the valves; it contained no clot. At its apex there was an opening which communicated with the pulmonary artery; it was oval in shape, four lines in length, with rounded edges. The right ventricle was a little dilated; its walls were hypertrophied. The heart weighed 15½ ounces.

* Clinical Medicine, Case cix.

In a case recorded by Professor Smith in the Dublin Medical Journal,* the pulmonary artery communicated with the aorta at its origin, by a small opening with rounded thickened edges. The aorta was dilated at the point where the opening existed. There was a loud blowing murmur heard with the first sound, and an intense purring murmur over the whole cardiac region. This purring murmur was heard in a case in which an aortic aneurism communicated with the right ventricle of the heart by two small rounded openings.† Dr. Bennett very justly observed that this tremor would not occur if the opening is large.

* Cited by Dr. Bennett.
† See Mr. Thurnam's very valuable paper.

CHAPTER VII.

HYPERTROPHY OF THE HEART.

ENLARGEMENT OF THE HEART.

This form of disease of the heart is met with under the following circumstances :—

1. Hypertrophy, without marked dilatation of the ventricles.

2. Dilatation of the ventricles without marked thickening of their walls.

3. With disease of the aortic, mitral, or pulmonary valves. In the last variety the symptoms are referable to the valves, the hypertrophy being secondary. See *Disease of the Valves.*

Hypertrophy is very common in the colony. It seems to be in some measure due to the muscular tissue of the heart being softer and less tenacious than in England; it is unable to bear any extra strain upon it, and the cavity of the ventricle dilates with or without an increase in the thickness of its walls. In strong, full-blooded men, the walls of the ventricle increase in thickness, with more or less dilatation of the ventricular cavity; but in the delicate, and those with soft muscles, the walls increase but little in thickness, while the cavity may be from one-half to twice its usual size. In the latter, the disease may assume, in the course of time, more or less the symptoms of flabby heart; the same may occur in the former, if the muscular system, by long rest, and the exhibition of digitalis, iodide of potash, mercury, and low diet.

It is often difficult to trace the changes which a case of hypertrophy undergoes under these circumstances. If an unfavourable opinion is given in a case of this kind, or if benefit is not received so quickly as the patient expects, other assistance is sought.

Hypertrophy was met with in 54 out of the 310 cases which applied at the institution; in 29 of the number, without disease of the valves, and in 25, with disease either of the aortic or mitral or of both valves. Obstruction of the pulmonary valves, or of the pulmonary artery, will produce hypertrophy of the walls of the right ventricle, or dilatation of its cavity. This is a change but rarely observed.

Influence of sex, age, and occupation, in predisposing to hypertrophy. The sex, age, and occupation, were noted in 42 cases; of this number, only 9 were females, to 33 males. The youngest of the females were 10 and 12 years of age, they went long distances to school, and were not very well nourished. Two of the others were 17 and 20 years of age; one had disease of the aortic valves, the result of rheumatism; the other chronic bronchitis, with asthma. They were domestic servants, and worked hard. A fifth was 22 years of age; she was pregnant, and had worked very hard as a servant. The mitral and aortic valves were diseased, the result of acute rheumatism. The sixth case was 33 years of age; the seventh was 40; the eighth 47, and the ninth 52. All these patients had worked hard at washing and mangling. The heavy strain on the heart in turning the mangle seemed to be the cause. In some of the cases, drink had to do with hastening the occurrence of the disease. It always hastened the fatal result.

The youngest of the 33 males was 13 ; he was delicate ; his father had died from disease of the heart: of two others under 20 years age (16 and 18), one was a lithographic printer, the other worked a heavy printing press. A fourth, a baker, was accustomed to carry heavy sacks of flour on his back. The next was a miner, aged 28.

It is more generally met with from the 25th to the 35th year, for 13, or nearly one-half of the 30 cases were between these years ; from 30 to 40, the liability was as great as from 40 to 50; but after 50, it diminished. Of four cases, one was 51 ; a second 52 ; a third 60, and a fourth 63.

The following table will show the relative liability at different ages, in the 31 cases :—

Before	20 years	... ...	3
„	25 „	... ...	0
From	25 to 30 years (inclusive) ...		7
„	30 to 35 „ „ ...		6
„	35 to 40 „ „ ...		3
„	40 to 50 „ „ ...		9
Beyond 50 years	... ...		3

It may be met with occasionally very early in life, but it is difficult to say from what cause.

Lads from 13 to 18 are very liable to attacks of palpitation of the heart on exertion. They can be traced to over exertion either at cricket, rowing, or running long distances, lifting heavy weights, or using the arms at work, either as saddlers, shoemakers, or blacksmiths. The attacks at first, only occur when the exertion is more than usually severe, but in time, if the strain is kept up, the palpitation is severer, the slightest exertion bringing it on, and with it the heart increases in size.

The first race born of European parents here are, I think, of larger growth than the second generation. Their growth is very rapid; they spring at once from childhood to manhood. Their blood does not appear to be so rich in fibrine; the clot is less firm than in the inhabitants of a climate like England; the muscles seem to be in a similar state. This rapid growth of the body is followed by exhaustion; they become prematurely aged. The average temperature of the body and of the breath is higher here than in England.*

The occupation followed exerts a great influence in producing the disease. Of 33 cases in which the occupation was noted—16 were labourers; 6 miners working in deep shafts, and in impure air; 2 printers, working heavy presses; 2 working engineers, and 2 blacksmiths.

The remaining 7 cases occurred in men following laborious occupations—shipwright, fireman on board of a steamer, sawyer, baker carrying heavy sacks of flour, painter, messenger, and slaughterman.

---

* The examination of the temperature of the breath in health, after sleep, food, and stimulants, and in disease, is a field in which no one has hitherto laboured. It promises to be of far greater value than the examination of the temperature of the surface of the body. In exhausting diseases it will be found to sink gradually; before death, when fluid exists in the bronchial tubes, it will sink below the temperature of the air, while in fever, inflammation, and after stimulants, it will rise. I use a piece of perforated cedar or pine of the shape of a respirator, with a small thermometer inserted so that the breath can pass freely around it during expiration; inspiration being performed by the nose. By means of a small mirror, the patient can note the temperature at different times of the day and under different ci. cumstances.

I have heard Bouillaud, in his lectures, speak of hypertrophy, as if it sometimes occurred as an acute disease. It may no doubt occur suddenly, as in the case of the painter, whose history will be found further on; but in these cases, the walls of the ventricle give, the cavity dilates, the patient will have palpitation, and the heart will be increased in size, but the pulse will be generally more feeble than in hypertrophy of the walls of the ventricle, without marked dilatation of the cavity, unless attacks of palpitation occur, then the pulse may become fuller; but it is never persistently quick and hard.

As a rule, hypertrophy is chronic and progressive, rather than acute. In tracing the history of hypertrophy without marked dilatation of the left ventricle, it will be found that the patient in full health works harder than usual, and is compelled during the effort made, to inflate and fix the chest; in doing this, the opening through which the aorta passes in the diaphragm is narrowed, and more or less pressure is made on the heart. The result of this is that if the aorta is unable to bear the strain, an aneurism will form, from its coats yielding, or the heart is called upon to use an extra amount of force to overcome the obstruction, and either the walls of the left ventricle increase in thickness and density, without any marked increase in the size of the cavity of the ventricle, or if the muscular substance of the heart is soft and flabby, or fatty, the walls yield, and the cavity of the ventricle increases in size.

COMPLICATIONS.—Of 29 cases in which the valves were not implicated, in 13 of the number, there were symptoms referable to the lungs—either consumption, pulmonary emphysema, bronchitis, either acute or chronic, or acute congestion of the lungs. In the last, the liver and the kidneys were apt to become implicated, but the urine rarely became albuminous, unless the patient had drunk heavily.

In two of the remaining cases, the pericardium adhered to the heart; in another case the aorta was dilated, and in a fourth it was ossified; in a fifth, there was a large round clot in the left ventricle—an unusual occurrence. (See case, page 140.)

In another case the patient was labouring under secondary syphilis. He said that he had never suffered from palpitation until the secondary symptoms had appeared. He did not seem to have worked very hard.

In the case of a man aged 63 (see case, page 143), symptoms of insanity set in, and while they existed the palpitation disappeared, but it returned as soon as he became sane.

Another case in which the mind became affected, has come under my notice. The patient, a chemist about 26 years of age, seen on the 18th September last, with Dr. Gregory. He had been taken a week before with acute rheumatism, accompanied with palpitation of the heart and difficulty of breathing. There was a loud, harsh murmur with the first sound, over the mitral and aortic valves; the second sounds were parchment like, but there was no marked regurgitation; the pulse was 120, full, hard, and regular. There was extensive dulness over the region of the heart. The right wrist was swollen and painful, but his chief complaint was of fulness in the throat. His uvula was larger than usual, but it was not swollen. It was touched with nitrate of silver, and considerable relief followed. In a few days he had a very distinct attack of diphtheria. It yielded to iodide of potash, quinine, and large doses of gum guiacum,* in a very few days.

After the throat got well, he had several attacks referable to the throat, not unlike hysteria, and violent attacks of palpitation of a choreic character. On October the 5th, he was much better; pulse 108; heart quieter. His aspect was peculiar. He sat up in bed with his eyes fixed and his eyebrows knitted. Has passed his urine in bed the last few days. The hysterical attacks ultimately passed into mania. He was then removed into the hospital, where he died about the middle of March. The heart was hypertrophied, it weighed 36 ounces (an unusual weight). Its right side was greatly dilated, especially the auricle. The mitral valve was thickened from rough calcareous deposit, and its opening narrowed, admitting only one finger; the aortic valves were thickened; calcareous deposit existed near them; one valve was perforated. The pericardium adhered generally to the heart. The lungs were generally adherent; they were congested. The liver was markedly cirrhosed, but it was recent. Spleen small; kidneys large and firm, but not granular. The brain was generally softened, especially about the fornix and posterior crura; the cerebellum was remarkably flaccid. There was a large quantity of fluid on the abdomen. He had been suffering for several years from disease of the heart. When seen by me, he had just been married. Some trouble had preyed heavily on his mind. The large size of the heart in this case was peculiar. In two cases

---

* I have used the gum guiacum in diptheria here, with very great benefit. If the disease is not unusually virulent, and the patient is removed from the house to a healthier locality, it rarely fails to check its progress in from forty-eight to seventy-two hours. Turpentine is a remedy which deserves a trial in the malignant form of diptheria.

which I had an opportunity of weighing the heart in simple hypertrophy, in one it was 28 ounces, in another 26. The patients were males. I have not found it to weigh more than from 16 to 20 ounces in the female.

Cirrhosis of the liver is a complication sometimes met with. This was the case in a female aged 48, stout and florid, 14 years in the colony, who applied at the Institution, June 11th, suffering from palpitation of the heart and difficulty of breathing on exertion, with pain and weight in the lower part of the chest after eating, followed in from two to three hours by vomiting of the food taken, mixed with bile. Her skin was jaundiced; liver large, extending from the nipple to two fingers' breadth below the false ribs. Heart's action increased in intensity, without any alteration in the valvular sounds; pulse full and strong, 108. The size of the heart could not be determined. She had had two children, and her menstrual discharge had ceased four years. She attributed her attack to exposure to wet and cold while washing and over exertion. It came on gradually. The jaundice disappeared under treatment, and the liver lessened in size, but with it dropsy set in. She was tapped several times. Death ensued in the month of December from exhaustion. The liver was reduced in size and cirrhosed; the kidneys were small. The heart was about twice the natural size; the walls of the left ventricle were thickened, but with the exception of the auricles, which were rather dilated, the ventricular cavities were not markedly increased in size. The occurrence of congestion of the lungs, liver and kidneys is an unfavourable complication; if acute from the suppression of the urine, death may take place in a very short time. In men who drink heavily and who are much exposed to the weather, dropsy generally sets in, first in the legs and then in the abdomen and chest. This was the case in a publican, seen with Dr. Stewart, between 40 and 50 years of age, from the neighbourhood of Eltham. The hypertrophy had existed some months; it had come on gradually from drink, exposure to the weather, and over exertion. His legs were swollen; urine scanty and loaded with lithiates; heart's action increased in intensity; its dulness extended over a large space, and blended with the left lobe of the liver. The liver extended below the false ribs. The posterior part of both lungs was congested. He was unable to lie down. The dropsy increased in intensity; his face became puffy and his lips congested. He died suddenly in a slight convulsive fit. It was with great difficulty that permission to examine the body could be obtained. The heart weighed 26 ounces; it was rounded, and very much enlarged, from thickening of the walls of the ventricles, which were distended with clots of blood. The lungs were very much congested; the liver and kidneys enlarged, and gorged with blood.* The abdomen and chest contained fluid. In early life he had suffered from spontaneous dislocation of the hip, but had recovered. I examined the capsular and the round ligament; they were natural.

The following are the best illustrations of this disease I have been able to collect :—

---

* When told that his case was hopeless, he said that Surgeon ——— would cure him for £20. He must therefore be considered to have died because he could not scrape that sum together. It is a common thing in Melbourne to see poor creatures in the last stages of consumption, heart disease, and cancer, deluded out of their money, and given a receipt in the following form :—" Received from ——— (£10, £15, and even £100), for which sum I undertake to cure him. Signed, ———, F.R.C.S.E."

Hypertrophy from over exertion, and extreme heat—Death.

A fireman, aged 30, strong and healthy-looking, applied, May 15th, at the institution, suffering for three months from pain in the chest, aggravated by eating. It was worse at night after going to bed, than in the morning and day-time. He has never drank hard. He attributes the attack to violent exertion, and the extreme heat to which he was exposed when replenishing the steamer's furnaces. He has never had rheumatism; pulse 120, full and strong; heart's dulness considerable, and its impulse strongly marked, with a slight murmur over the mitral and aortic valves. The aorta can be easily felt behind the sternum; liver natural; urine high-coloured, loaded with lithiates. He has great difficulty in walking fast, or up a hill. He can lie best on the left side. He complains more of the pain after food than of any other symptom. He improved under the administration of bromide of potash, digitalis, and ammonia.

He was lost sight of until June 20th; he had got gradually worse; the heart's impulse more strongly marked; pulse full and strong, 120; feet have a tendency to swell; urine scanty; food aggravates the pain in the region of the heart; is unable to lie down. Twelve leeches were applied to the chest, and the tincture of veratrium viridium given. These measures relieved the pain, and lessened the force of the heart's impulse, and the fulness of the pulse, for a few days.

June 30th.—He is unable to swallow unless he bends forward. He sleeps and takes his food kneeling, with the left side of the chest resting against a cushion on a sofa. He says that this is the only position he can take food (liquid) without pain, and sleep without dreaming that he is suffocating. If he attempts to swallow food when in the erect position, it is arrested opposite the lower part of the sternum, and a pain shoots down the right and left arms, and to the nipples. The pain is more severe in the left arm and left nipple than in the right arm and nipple. He thinks the veratrium viridium of service; it reduced the strength of the pulse, but not its frequency. During the next fortnight he got weaker; his legs swelled considerably and he became subject to severe attacks of pain in the lower part of the chest. They seemed to be of a neuralgic character, and were unattended by any increase of the heart's action, or of the pulse. From a fourth to half a grain of morphia relieved these attacks for a short time. The day before death, his face was dusky coloured, his lips bluish, and his nose pinched; but beyond being weaker, there was no very marked alteration in the symptoms. He had a fit in the night, and died suddenly.

It was with great difficulty that permission could be obtained to examine the chest. The heart was enormously enlarged; it weighed 28 ounces; the walls of both ventricles were greatly thickened, particularly those of the left one. The left ventricle was distended with clots of blood; the right one was empty. The heart looked, both in shape and appearance, like a bullock's. The aorta was unusually large; the pulmonary artery was also enlarged. The valves were healthy, with the exception of the liver, which was dense and bluish; the organs of the chest and abdomen were healthy.

In the next case a large dense clot formed in the left ventricle. The heart was not nearly so large as in the last one. I had no means of weighing it.

The patient, a florid-looking labourer, aged 32, fourteen years in the colony, applied at the institution on the 18th of September, suffering for two or three years with palpitation of the heart, and difficulty of breathing, which had become so severe as to prevent him working. He thinks that it was first caused by using a heavy, long-handled shovel. He has slight cough, and expectorates a little clear phlegm, which is occasionally tinged with blood. He has always been very sober, and enjoyed good health, with the exception of an attack of jaundice and dysentery, seven years before. The heart's action was very much increased; there was considerably more dulness than natural; pulse full and hard, but not increased in frequency. There was a slight murmur heard over the mitral valve, and some slight indications of pulmonary emphysema; liver was slightly enlarged, but there was no dropsy; urine natural. A blister was applied to the chest, and small doses of digitalis, with iodide of potash, ordered.

28th.—He has got cold; coughs and spits a good deal, and has passed a large quantity of blood from the bowels. Face pale, and ex-sanguine. There was marked congestion in the posterior part of both lungs; the pulse was very feeble and irregular; the heart was acting tumultuously, a churning sound having taken the place of the murmur before heard over the mitral valve. His abdomen was very tender, and he complained of considerable pain in it. In the evening his breathing got gradually difficult, and he died at 11 p.m.

The heart was very large; the walls of the ventricles were thickened, and when divided, they cut with a grating sound, as if the muscular fibres were intermixed with cartilage. All its cavities were empty, with the exception of the left ventricle, which contained a large, dense, fibrous clot, of the size and shape of a small hen's egg. The lungs were emphysematous in front, but behind they were congested. The liver was very dense; the kidneys the same; both were congested. The peritoneum was also congested; there was about a pint of fluid in the cavity of the abdomen.

In the next case, the patient was young :—

Symptoms referable to the Stomach—Hypertrophy of the Heart—Dropsy—Death.

A male aged 13, a native, of a scrofulous aspect. His father had died from disease of the heart. He applied to the institution Nov. 6th. He had been ill for four months, suffering from pain in the region of the kidneys, and vomiting of bilious fluid and water, sometimes saltish, sometimes insipid. His mouth was inundated with water, both during the day and the night. At first his scrotum was swollen, but now his feet and legs as high as the knees. He is unable to lie down at night, from palpitation of the heart, and difficulty of breathing. The action of the heart was greatly increased; there was considerable dulness on percussion, but no alteration in the sounds; pulse 70, rather full; liver dulness increased; kidneys free from pain and tenderness; urine scanty, and high-coloured, but not albuminous. The attack was excited four months before, by playing at football, and getting wet when heated, and remaining for some time in his wet clothes. By encouraging the action of the skin with diuretics and strict rest, he improved considerably, but did not recover his former health. He remained a patient up to February, his state varying, being sometimes better, sometimes worse. In the first week of that month, he got cold, and over exerted himself. His abdomen began to swell, and his lungs became congested, the liver enlarged, and the heart's action increased in loudness; pulse rose to 110; urine became scantier, but it was free from albumen. He could not lie down. The vomiting, and the escape of water from the mouth had ceased for some time, but they were replaced by watery diarrhœa. His appetite was very bad. Various plans of treatment were tried without any effect. He got gradually weaker, and died in a fit at the end of the month.

The heart was as large as a full grown adult's; it was rounded. The left and right ventricles were dilated; they both contained a little pink blood. The walls of the left ventricle were slightly thickened. Both the lungs were slightly congested in their posterior parts; there was several ounces of serum in the cavity of the chest, and about an ounce in the pericardium. The abdomen contained about four pints of serum; the liver and kidneys were enlarged, but not markedly congested. The pancreas was larger than usual, and of a deep, venous hue.

In the next case, the heart dilated suddenly, during violent exertion.

Hypertrophy, with attacks of Chorea of Heart—Insanity— Cessation of the Heart Symptoms, which returned before Death.

A painter aged 63, fifteen years in the colony, applied at the institution November 18th, suffering from difficulty of breathing, with palpitation of the heart on exertion, and inability to lie down at night. He gets distinct attacks of palpitatation towards morning, from 2 to 4 a.m., and they last from one to two hours. He has never had rheumatism, or drank hard. He has been ill three months, and states that the palpitation was first excited by walking fast up a long hill, while carrying a heavy basket on his back. He felt his " heart give way," when near the top, and was obliged to sit down. He walked home with some difficulty. The slightest exertion brings on the difficulty of breathing, and palpitation. The heart's dulness was considerable; its impulse marked with a slight murmur in the mitral and aortic valves, particularly in the latter; pulse 112, full and regular. There was no arcus senilis, and his sight had not failed for some years. He can lie best on the left side, and on the back, but not on the right one for more than a few minutes. By strict rest, and the exhibition of digitalis, with ammonia and bromide of potash, leeching the chest, followed by the application of a belladonna and opium plaster, he was so much better by the 28th of December, as to be able to lie on either side, and the choreic attacks had ceased to occur; the pulse had fallen to 90, and the heart's dulness had lessened in extent. Shortly after this, from walking on a hot-wind day, all the symptoms returned. Soon after this his mind gave way, and he became subject to a mild kind of delirium; for ten or twelve weeks during its existence, the heart ceased to trouble him. The organ remained large; the pulse was 120, but not so full as it had been. When the delirium ceased, the heart symptoms became again troublesome. He died on the 2nd April, worn out by loss of sleep. He had suffered very much from mental anxiety.

The heart was very much enlarged ; it was nearly round. All the cavities were very much dilated—the left ventricle to nearly three times its natural size. The walls of the ventricles were not much thicker than natural; they were free from fatty deposit The valves were natural; the aorta the same, but it seemed somewhat increased in size. The other organs were healthy. The head was not allowed to be examined.

Treatment.—In young lads or men, *strict rest* for several months, doing nothing that can excite the heart, will enable it to recover its natural size. The diet should be very light and unstimulating; digitalis and belladonna, with bromide or iodide of potash, and very minute doses of calomel, may be given. Leeches are of great service when the heart's action is increased and the pulse full and strong. It must be remembered that patients do not bear depletion so well here as in England. The object must be to bring the pulse to as near its natural standard as possible, and keep it so. If the strength is too much reduced, the pulse has a tendency to increase in frequency, and then attacks of palpitation are liable to occur. When the patient is anæmic, it is well to combine, and particularly when the ventricles are dilated, the digitalis with the solution of perchloride of iron; from three to five minims of each being given three times in the twenty-four hours. In some cases the veratuum viridium is of service in reducing the circulation, but it is better adapted for hospital, from its action being more easily watched, than in private and out-of-door practice. The ice-bag applied to the region of the heart, is often of great service when the attacks of palpitation are violent. Some patients feel greatly relieved by wearing a belladonian and opium plaster.

When congestion of the lungs, liver, and kidneys, occurs, salines with mercury and podophylline are often of service. The action of the skin should be increased by packing the body in a

L

hot wet sheet.*   When dropsy occurs, the same measures may
be had recourse to.   Elaterium, although giving great relief to
the symptoms, is not very well borne here.   It reduces the
strength too much.

* I have used the hot wet sheet here instead of the cold one. It does not
chill the patient. The following is the way to employ it :—Dip a sheet in hot
water, wring it out as dry as possible, and envelope the patient in it. He
should be enveloped with three or four blankets, and two or three bottles filled
with hot water should be placed on each side. Copious perspiration is soon
excited. When this is thought to be sufficient, the body is sponged with
tepid vinegar and water. To avoid cold, the patient should be put to bed for
a time.

CHAPTER VIII.

FATTY DISEASE OF HEART AND FLABBY HEART.

1. FATTY DISEASE OF HEART.—Fatty disease of heart is involved in considerable obscurity, its existence being generally only discovered after death. During life the pulse will vary very much. In one person it will be full and weak when quiet, becoming feeble and irregular on exertion, with more or less palpitation, or "fluttering," or failing of the heart's action. In another, the pulse when lying down is natural, but in the erect position or on exertion, it alters—becomes quick and feeble, or soft and irregular. The heart's action will vary in some cases; it acts more or less violently, but not with the force observed in nervous palpitation or hypertrophy, but with a want of impulse difficult to describe, but which, when once heard, cannot be easily forgotten. In others the heart's impulse seems to be confined to the point where the ascending vena cava passes through the diaphragm. In the cases of fatty degeneration which I have met with here, it has occurred under the following circumstances :—1st. With great deposit of fat under the skin and about the viscera. 2nd. With fatty degeneration of the liver, either without or with deposit of fat under the skin and about the viscera, in men and women who drink largely either of beer or spirits. If they are stout they are generally great eaters ; if thin they live nearly altogether on drink. The aspect of the former, class of patients is peculiar ; they may be more or less florid, or

pale and pasty-looking, the muscles of the arms and thighs more
or less soft to the feel. 3rd. With fatty degeneration of the
kidneys. This generally occurs in connection with fatty degeneration of the liver, but not so frequently or so well marked here
as in England. In these cases the patients had been very much
exposed to the weather, often wanting food, and caring little for
it as long as they could get drink ; some of them drinking from
one to two bottles of spirits in the course of the twenty-four
hours. There is a fourth form of fatty degeneration, the result of
a change which inflammatory deposit between the heart and the
pericardium seems to undergo in time. In a case in which there
was a layer of fat larger in size than a seven-shilling piece on the
surface of the heart, the muscular substance for a considerable
depth was affected with fatty degeneration. The patient, a male,
50 years of age, had died of apoplexy. There was considerable
deposit of fat about the body.

INFLUENCE OF SEX AND AGE.—Of 310 cases of diseases of the
heart, in 13 there was fatty degeneration. Since this table was
compiled, eight more cases have come under my notice at the
Institution.

Females are very much more liable to the disease than males,
for out of 21 cases, 15 of the number were females.

With reference to the ages of the females, all were beyond 40
years of age and under 60, only one being 40 and one 60. From
45 to 50 and 50 to 58 being the periods when it is most liable to
occur. Of the 15 cases—

6 were between	...	...	...	40 and 45
3 „ „	...	...	...	45 „ 50
6 „ „	...	...	...	50 „ 60

It would seem that on the cessation of the menstrual discharge there is a strong tendency to the deposit of fat under the skin and about the abdominal organs. It is particularly liable to occur in females who have been some years in the colony, and who have worked hard in their youth, whether they have had large families or not. With but very few exceptions the patients were unusually stout. When they were not they were great drinkers.

Of the six males—

    3 were between   ...   .   43 and 50

    3 ,,   ,, ...   50 ,, 55

Both males and females followed sedentary occupations, or took but little exercise, eat freely, and drank more or less.

It would seem that when either sex show a tendency to become stout, the muscular substance of the heart and the other muscles become more or less liable to fatty degeneration. In a woman, about 40 years of age, who died in the hospital, the upper part of both lungs were fatty; the heart, liver and kidneys participated. She had had inflammation of the lungs two years before admission. She had remained in bed since the attack and grown stouter than she was before. Her skin was of a waxy hue. The same colour of skin existed in another case. In the former case, death ensued from apoplexy; in the latter, from diarrhœa.

SYMPTOMS.—The symptoms are extremely uncertain. They will be very much influenced by the extent of the degeneration and by the amount of dilatation of the ventricular cavities. If the degeneration is not extensive, but little complaint will be made, but if with it the ventricles, particularly the left one, is dilated, then they will be more severe. Very extensive fatty degeneration is often found after death, its existence during life being unsuspected, as if the organ remained capable of doing the work it was called upon to perform, until from some cause its powers were overtaxed and death ensued. Whenever a woman of sedentary habits, beyond forty years of age, begins to get stout, and with it complains of difficulty of breathing, with palpitation of the heart on exertion, and attacks of palpitation or fluttering at night occurring after the first sleep, with a tendency to faint, the pulse being either full and soft and easily compressible, or quick, feeble and compressible, fatty degeneration may be suspected. The rest which the difficulty in walking inclines the patient to take encourages a still further development of fat, the appetite being often so good that the patients, from a habit which they find difficult to control, both eat and drink more than they should do, and increase the disease. The debility, sinking at the pit of the stomach, and palpitation or fluttering of the heart, which long abstinence from food excites, compels them to take food or stimulants oftener than they otherwise would do. The heart's sounds are feeble and without power. It is not possible in stout females with large breasts to determine the size of the organ.

Extreme fatness is the usual attendant on fatty degeneration, but in drunkards this is not always the case, particularly when they live on drink. They may be thin, but the liver, unless dropsy and cirrhoses has set in, will be found more or less enlarged and fatty; the kidneys often participate. The patients rarely complain in these cases of heart disease; they seldom walk far, or if they do are easily fatigued and need a stimulant to keep them up. The pulse in these cases is usually quick and feeble. The skin is pale and waxy. Fat may sometimes be discovered in their motions, and in the urine fat granules and sometimes albumen.

These patients are particularly liable to attacks of bleeding from the bowels. They may either sink from this, from diarrhœa, vomiting, or exhaustion. In several of these cases, in which I have had an opportunity of making post mortem examinations, the heart was not enlarged, the ventricular cavities were small, and either empty or containing but a small quantity of thin blood; their walls were often extremely fat; the liver and kidneys more or less so.

In the cases of extreme stoutness, death usually occurred either at night, the patients suddenly complaining of difficulty of breathing and died before they could be seen by a medical man, or while walking fast they fall to the ground in a fainting state. Instead of being allowed to lie they were raised into a sitting position, which interfered with the heart's action, and they not infrequently died. Or after a heavy meal, they ran a short distance, became breathless, and died in a few minutes; or from the vessels of the brain being in a state of fatty degeneration, they were either seized with paralysis or apoplexy.

The post mortem appearances in these cases were as follows :— The heart was larger than natural, the walls of the ventricles soft and fatty, their cavities either distended with blood or empty. These two states seem to depend on whether the heart had power to empty itself when full, or when empty able to receive the blood. Both these states may depend on obstruction. In a man 50 years of age, who after a long fast, and walking a long distance, eat a hearty supper of potatoes and meat, and drank a large quantity of tea, two hours after going to bed he woke up, complained of difficulty of breathing, and died in a few minutes. The heart was distended with blood. The same state existed in a

female aged 58, who died suddenly from apoplexy. In a woman
aged 45, who while running after a news boy to whom she had
given half a sovereign instead of sixpence, the heart was quite
empty. Her stomach contained two-thirds of a wash-hand basin
full of peas, potatoes, and ham in an undigested state. In a stout
man 55 years of age, a great eater and drinker, suffering from
enlargement of the liver, raising him, just after he had eaten a full
meal, suddenly in bed, caused death. The heart was empty. In
drunkards, when they are much emaciated, the cavities are often
empty, or they contain only a little thin blood.

COMPLICATION.—Two of the females were liable to epilepsy;
three had had attacks of paralysis, and one of the three died of
apoplexy.

Valvular disease was very rarely observed;—but few of the
patients (only two) had had rheumatism. In drunkards hæmorrhage from the bowels was sometimes observed, and occasionally
diarrhœa. If persistent, they generally died. Very little loss
of blood or slight diarrhœa was often sufficient to destroy life.

DIAGNOSIS.—The only disease with which fatty degeneration is
likely to be confounded is flabby heart. This is however a disease of
early rather than of late life. The patients are thin and pale, work
in confined situations, and use their arms a great deal. The cases
which have fallen under my notice were either printers or lithographic printers, working heavy presses. I have seen this disease
occur in men from 25 to 30 years of age, who after following
sedentary occupations for some years, take to out-of-door work.
Persons suffering from fatty disease may from the left ventricle
dilating present symptoms of flabby heart. A man 43 years of
age, who had been under my care for some time suffering from

symptoms of fatty disease of the heart, which was relieved by treatment, after carrying a heavy load on his back for some distance, felt his heart give; he became faint, and was obliged to lie down. From this time the attacks of palpitation became very severe and more liable to occur at night than before. He was not able to walk far without great risk of fainting. The heart's sounds were weaker than before, and the pulse very soft and compressible. The area of the heart's dulness had extended. He again improved under the use of oxide of zinc, with small doses of morphia and nitric acid, and strict rest. At the end of a month he was so much better that he ceased to attend.

TERMINATION.—The tendency seems to be to end fatally, when the patients are beyond 50, and particularly if it is complicated with paralysis or fatty degeneration of the liver and kidneys. There is a great risk in its ending fatally if the patient continues to get stouter, and if with this the symptoms referable to the heart increase in severity. An overdose of strong opening medicine, particularly of blue pill and black draught, may cause so much exhaustion as to produce death. Chloroform, hydrate of chloral, digitalis in large doses, veratrium viridium, are all liable to cause it to end fatally. Calomel when very freely given may, by exhausting the heart's power, cause a fatal result. In drunkards the disease is nearly certain to end fatally. Hæmorrhage from the nose sometimes occurs. It is often very difficult to check, unless the nostrils are plugged. Great exhaustion will follow, and if the patient's strength is not well supported by stimulants and beef tea, attacks of fainting will be liable to occur. Suckling acts very injuriously in this disease. In the case of a very stout, pallid and flabby female, aged 40, 15 years in the colony, who had been a patient at the Institution several times during pregnancy, suffering from symptoms of fatty heart, suckling her child aggravated all the symptoms, and brought on attacks of fainting in the mornings from 10 to 11, and "fluttering" of the heart in the night, which threatened to destroy her. Her pulse was 120, soft and feeble. She had been liable to attacks of bleeding from the bowels and nose. She weaned the child, and by rest and tonics recovered. Her heart remained weak. She had a great tendency to make fat. In another case, the patient, 46 years of age, stout and florid, had been complaining of difficulty of breathing for three years. Six weeks before seen, while carrying her child, eight months old, some distance, her heart began to beat, and since this any exertion caused palpitation. She is liable to palpitation soon after going to bed; pulse 115, very feeble; heart's action feeble. On weaning the child, and by rest and tonics, she soon recovered; as in the other case, the heart remained weak for some time.

TREATMENT.—Most of the cases derive benefit from the acids, particularly the nitric, or from iron and quinine combined, when they are liable to "hot flushes," with sulphuric acid. Occasionally the iron and quinine is too stimulating. This was particularly the case in two instances, the patients complaining that it aggravated the symptoms. There did not seem to be any very marked reason for this. Some patients seemed to be benefited by small doses of digitalis, given with ammonia, and others by oxide of zinc, with small doses of morphia and strychnine.

Diet and exercise are most important in the treatment. Both should be carefully regulated to suit the patient's strength; the former should not be reduced too suddenly, or the latter carried so far as to overtax the strength or the powers of the heart. It is difficult to lay down any rules as to diet, for some females will get stout on what would starve others. Meat should be taken only once a day, and then not more than four ounces with one kind of vegetable and dry bread with water. For breakfast and tea, tea without sugar or milk, or plain milk and water with a biscuit or a piece of dry bread. This diet should not be too suddenly adopted, as it is liable to cause so much exhaustion that the attacks of fainting will become more frequent and much worse. The stomach should never be left too long empty. In exercise, the same rules should be observed as in diet. The patient should be impressed with the necessity of walking a certain distance every day, increasing it as the strength permits. The Turkish or hot air bath is of great service in some cases, but the use of either needs care, for some patients do not bear being sweated too freely at first. Neither must be considered as a substitute for diet and exercise. Patients should be told to weigh themselves regularly, and regulate their diet and exercise in such a manner that they reduce their weight about one or two pounds every seven or ten days. If they do this, their healths will not suffer, as it is apt to do when they are reduced suddenly.

2. FLABBY OR FLACCID HEART.—In this disease the heart is enlarged ; but, unlike hypertrophy, the increase in size is due to the dilatation of the ventricles and auricles.

In the cases which I have had an opportunity of examining after death, the heart was found lying at the bottom of the pericardium, like an empty bag, of a pale or dark slate colour. It does not seem to participate in the rigor mortis, for in a man who died from the effects of strychnine, it laid flat and empty at the bottom of the pericardium, although the muscles of the legs, arms, back, and abdomen were rigidly contracted. The valves were healthy; the lining membrane and the substance of the heart were pale. The heart was free from fat.

The heart, in very marked cases, seems to resemble a soft fibrous bag. This was particularly noticeable in a case reported in the chapter on Ossification of the Pulmonary Artery.

In many of its symptoms it bears a very close resemblance to fatty degeneration; there is the same feebleness of pulse, want of strength, and difficulty of exertion, from failure of the heart's action. The two diseases may occur together. Over exertion in fatty alteration, if the walls of the ventricles are weak, will cause them to yield, their cavities remaining permanently dilated.

Ossification, or obstruction of the pulmonary, aortic, or the auriculo-ventricular opening will, if the walls of the heart are weak, cause the same alteration. This was strongly marked in the case of ossification of the pulmonary artery. *See diseases of this vessel.*

Patients suffering from this disease are very liable to a kind of *asthma* on exertion, but it is unattended by either cough or expectoration. The attacks occur at first at long intervals, but over-exertion causes them to increase in frequency; the pulse is soft and compressible, ranging from 40 to 90. They are very liable to what they call " weak fits," particularly if excited, or if they go too long without food, and often from the same cause wake up in the night, and complain of a sense of sinking through the bed. The heart flutters, or ceases to act, during these attacks; the surface of the body becomes cold, or covered with perspiration. They feel as if about to die.

A thin, delicate, medical man who had suffered from consumptive symptoms in England, on coming out here, walked a great deal, and worked hard mentally. His breathing while walking would become suddenly difficult; the action of the heart stop for several seconds, then flutter, and if he did not grasp something, he was in great risk of falling. These attacks were very liable to occur when returning home in the morning, after being up all night. Any mental shock produced nearly the same effect, but fluttering of the heart preceded the difficulty of breathing, and the tendency to faint. By riding on horseback, and taking stimulants more freely than he had been accustomed to do, he quite recovered.

In cases of hypertrophy, free depletion, mercury given for some time, with low diet, and a very long rest until the muscles became flabby, and the skin pallid, over-exertion may throw such a strain on the aorta that either it or the ventricle dilates.

Chlorotic girls, pallid young men who have just attained their full growth, if addicted to smoking, self-abuse, excessive intimacy, and late hours, if compelled to work hard—are liable to this disease.

Of the seven cases which applied at the Institution, four were males, their ages ranged from 25 to 30 years. Two of the females from being machinists, had become general servants; both were chlorotic; and the third, a female aged 33, stout and flabby, had been obliged to exert herself more than she had been accustomed to do, on the death of her husband.

When flabby heart occurs in young men—printers, or lithographers, it is from the muscular effort made when the breath is held while working the press, the heart is called upon to use more force than it is able to bear, its muscular tissue being weak, yields. These patients are pale and emaciated. The same exertion in muscular patients of sanguine habits would, if it affected the heart, produce hypertrophy.

It is difficult to diagnose this disease from dilatation of the aorta or the pulmonary artery, and fatty degeneration of the heart. It often co-exists with one or other of these diseases. See *Dilatation of the Aorta and Pulmonary Artery, and Fatty Degeneration.*

Both dilatation of the aorta and of the pulmonary artery may produce the form, of asthma described.

The increased size of the heart, the feeble pulsations, and the difficulty in walking any distance, or at a quick pace, the pallid state of the skin, soft muscles, and the absence of fatty deposit, are indicative of this form of the disease.

The following additional cases may help to illustrate this disease :—

A widow aged 33, stout and pale, had been for nearly a year a patient at the Institution, suffering from symptoms of flabby heart. The death of her husband compelled her to work very much harder than she had been accustomed to do. Her menstrual discharge, which had always been profuse, became still more so, lasting from fifteen to twenty-one days. She had had nine children in about twelve years ; her labours were always attended with more or less flooding. On the 19th of December, I was requested to see her, as she had been suffering for a longer period than usual from the discharge. By plugging the vagina, and giving her large doses of gallic acid, the flooding was checked. The discharge came on again at the end of ten days, and was allowed to continue for a fortnight, before she sought assistance. The plug was again used, but although warned not to allow it to come away, she removed the bandage. In the morning she attempted to get out of bed to the close stool. She felt very faint, and complained of sinking at the pit of the stomach, and fluttering of the heart. Brandy was given. She was lifted into bed with great difficulty! she expired as they laid her head on the pillow.

The body was not very fat, the chief deposit being under the skin. The lungs were pale; the pericardium contained a little fluid; the heart laid quite flat in the pericardium, scarcely filling one-half of it. Its cavities were quite empty, and enlarged; the walls were pale and spongy, but not markedly fatty. There was a little deposit of fat on its surface. The muscles of the body were pale and soft, and easily broken up. The uterus was large and flabby; it contained a soft clot of the size and shape of a pear. The head was not allowed to be examined. The excessive loss of blood after labour, or during a miscarriage, the too free use of strong drastic medicines, and over-suckling, or over-exertion, when the system is exhausted by long fasting or the want of its accustomed stimulant, may cause flabby heart to prove fatal. The too free use of chloroform, or a very large dose of chloral may do the same.

A cachetic woman about 30 years of age, who had suffered for several years from "trouble about the heart," died suddenly, while attempting to sit up in bed, a few hours after flooding. The heart was empty and flabby; the womb was very large and soft, it contained a large clot. The brain was very pale. The veins of the chest and the abdomen, and the sinuses of the brain were gorged with blood. The medical man who had confined her was accused of neglect.   The abdomen was well bandaged.   The nurse was perhaps to blame for allowing her to sit up.

TREATMENT.—The patient should be strictly cautioned not to do anything which can bring on an attack, but should not be allowed to sink into a state of idleness.   Sponging the chest and body with cold water, or the cold bath, is generally of service. Iron and quinine, with a minute quantity of morphia or ammonia, with calumbæ, and decoction of cinchonæ, should be given.   Two or three glasses of good red wine may be taken with lunch, dinner, and supper, and beef tea in the course of the night.   Care should be taken not to let the patient go too long without food.   He should, if going any distance, either carry a well-boiled egg or a biscuit, or some brandy and beef tea.   Spirits rarely give more than temporary relief, if taken without nourishment.   Smoking should be prohibited.

CHAPTER IX.

1. RUPTURE OF THE HEART. 2. RUPTURE OF THE VALVES

3. CANCER. 4. HYDATIDS. 5. ANGINA PECTORIS.

1. RUPTURE OF THE HEART.—No part of the heart is exempt from rupture. The walls of the auricles or of the ventricles may give way, the blood escaping into the pericardium. Occasionally the interventricular septum yields, and the venous and arterial blood becomes mixed. The heart may give way from ulceration, softening, fatty degeneration, or in consequence of an abscess in the walls of the ventricle, rendering the tissues so thin that they are unable to bear any extra strain. The presence of bony deposits on the surface of the heart or in its substance may produce some alteration in its vicinity, which may render the parts unable to bear the pressure of the blood, particularly if the patient exerts himself more than usual. This was the case in a man, aged 60, who suffered from severe pain in the region of the heart, on the slightest exertion. The exertion did not increase the quickness of the pulse. The radial vessels felt hard and unyielding. The heart's sounds were natural, but feeble. He died after 36 hours' illness, the chief symptoms of which were extreme feebleness following a severer attack of pain than usual. There was a layer of calcareous substance larger than a two-shilling piece, irregular in shape, and penetrating deeply into the substance of the right ventricle. A fissure existed at the side of the deposit, and it

M

was through it the blood had escaped into the pericardium. The muscular substance was very soft in the vicinity of the fissure. There were calcareous deposits on the pericardium. The thoracic portion of the aorta was very much ossified.

Males are more liable than females, for of six cases which have fallen directly and indirectly under my notice, all were males; their ages ranged from 50 to 70 years. It is sometimes difficult to say what causes the rupture; violent exertion seems to have a great deal to do with causing partial rupture of the fibres, some alteration being excited that weakens the remaining part, which, under a very slight strain, gives way. Death may take place suddenly, the patient just crying out "My heart has given way," or, "Oh! my heart," if the opening is large; but if it is small, they may faint, recover slightly, but fall again into the same state, remaining in it until death ensues. This was the case in which a pouch formed in the walls of the ventricles. The patient, a stone-mason, 55 years of age, was under care for several weeks, suffering from pain in the region of the heart, particularly on exertion. He said that when lifting a very heavy stone, he felt something give way in his side near the nipple. The pain caused him to faint. He had never been able to work after it. There was nothing either in the pulse or the state of the heart to lead to the supposition that there was any severe lesion. As he had been addicted to drinking heavily at times, and neglecting his work, his friends rather doubted the truth of his statement.

About the end of the seventh week, while in bed, he complained suddenly of more pain than usual in the region of the heart. He had been to the closet just before the attack of pain came on, and had strained considerably in passing a hard motion. Half an hour after the attack, his face was pale, covered with cold sweat, skin cold, pulse nearly imperceptible. The heart's sounds could scarcely be heard, there was a marked increase in the size of the heart. Brandy and ammonia rallied him for a short time. He died at three a.m., about five hours from the time he was taken. In the left ventricle, close to the wall which divided the two ventricles, there was a small pouch about as large as a walnut, the outer wall of which was formed of muscular fibres and the pericardial membrane. Near the apex of the pouch, there was a small ragged opening which would about admit the point of a No. 2 catheter. The pericardium contained from one pint and a half to two pints of blood, partly clotted and partly fluid. The substance of the heart around the opening was very soft: it was generally fatty. The liver and kidneys were enlarged.

Dr. Moore, of Hotham, has met with a case of rupture of the heart. The patient, a man near 70 years of age, after walking upstairs, undressing, and getting into bed, uttered a cry, and immediately expired. There was a rupture in the posterior part of the left ventricle large enough to admit the point of the little finger. The pericardium contained a very large clot of blood. The valves were healthy, but the substance of the heart was so softened that the point of the finger could be easily driven through it. He had never complained of his heart.

In a patient in the hospital suffering from hydatids of the liver, the symptoms were not unlike those of the first case. After death, the pericardium was found distended with blood. There was a small aneurismal pouch immediately above the aortic valves; its apex had ulcerated. The cyst had been tapped, and he left the hospital quite well. He returned three weeks afterwards, on the 6th of July, complaining of pain in the back, between the shoulders, which he said had shifted to the chest. He had cough, and a little expectoration of thick, ropy mucus, and at times vomited up his food. The day before admission, he had had an attack of severe pain in the lower part of the chest, on the left side of the sternum. It came on suddenly while going up two flights of stairs; he vomited up his dinner, which he had taken a short time before, and became very faint. To-day he had another fainting fit, but it was not so severe or of such long duration as the one he had yesterday. When examined on the 7th, his pulse was very feeble, 108, equal in both wrists; his skin pale; heart's sounds feeble. The heart's dulness was increased somewhat in extent, but the valvular sounds, particularly the aortic, were natural. The symptoms seemed to indicate the occurrence of hæmorrhage, external to the hydatid cyst. The cyst had not increased in size. He remained very much in the same state until he died in a fainting fit, on the night of the 14th. The pericardium contained two pounds and a-half of blood, fluid and clotted. The pericardium was deeply dyed with the colouring matter of the blood. After considerable search, the source of the hæmorrhage was found to be the small opening in the pouch. This pouch was distended with dark and rather hard clots; it was lined with several layers of loose lymph. It seemed that they had closed the opening for several days—from the 6th to the 14th. There was a second sac on the opposite side to the other, but smaller in size; it was empty. Both sacs were immediately behind the aortic valves. The first one adhered to the pulmonary artery, the walls of which were very thin. In the right lobe of the liver there was a nearly empty hydatid cyst, the size of the two fists. Its upper border was imbedded in the liver; its lower one adhered to the gall bladder. Its external coat was fibrous, studded with rough, bony particles; its internal one was like condensed white of egg, and very easily torn. The fluid in the cyst was semi-purulent; it did not exceed three ounces. A quart of fluid was removed when he was in the hospital the first time.*

I have seen no case of rupture of the inter-ventricular septum. It produces symptoms resembling those met with where there is a communication formed between the aorta and the pulmonary artery. See *Diseases of the Pulmonary Artery.*

Louis collected a number of valuable observations from Morgagni, Moreau de la Sarthe, Bouillaud, Corvisart, Breschet, and Bertin, Hunter, and several others in which communications occurred between the different cavities of the heart.†

2. RUPTURE OF THE VALVES.—This is an accident rarely met with in the colony. In the chapter on hypertrophy a case will be found reported in which there was an opening in the aortic valve. In a man who died in the hospital from dropsy and effusion into the chest, two of the aortic valves were detached (one for one-third, the other for less than one-half) from the aorta. The heart was enlarged, and the aorta was ossified, and more dilated than usual. The patient, a digger aged 40, died on the second day of his admission.

3. CANCER OF THE HEART.—Lebert, in his valuable work

* Hydatids are extremely common in the colony. They can be usually traced to eating undercooked hydatid mutton, drinking water from ponds frequented by sheep and dogs, and eating watercress growing in districts where hydatids abound.

† Memoires ou Recherches Anatomica-Pathologiques, 301.

on cancer, does not mention cancer of the heart. From cancer being very common in the colony, very much more so than in Europe, cancer of the heart or pericardium may be expected to be met with. Walsh says that cancer is unknown in Australia. I do not know on what authority he makes this statement. Hope says that not more than a dozen cases had been met with when he wrote.* He considers that it is developed in two forms (1), *isolated,* and (2), *interstitial infiltration,* and that they rarely occur without similar productions in other organs, especially the lungs. These observations apply to the disease as observed in the colony.

I have seen two cases. The patients were men, and both attributed the disease, the true nature of which was only discovered after death, to injuries of the chest. Both the patients were about 40 and 45 years of age; they had drunk heavily. In one of the cases, the patient, a builder residing in Richmond, stated that after a fall from a roof, pain in the chest set in. When seen three months after the accident, the pain was very severe at night; it lasted from two to four or five hours. There was dulness on percussion, over the sternum, below the third costal cartilage; it was much more pronounced on the left than on the right side. This dulness occupied an irregular space about seven fingers broad, and seven deep. The heart's sounds were very obscure, they quite disappeared at the end of a month. The aortic sounds were readily heard from the upper border of the third costal cartilage, but both the aortic and mitral valvular sounds were inaudible. Altering the position of his body did not, as in pericardial effusion, render them more prominent. The pulse was

* Page 365.

natural, but feeble. His aspect presented no signs of cancer His face was pale, but his body was well nourished. The sternum and the costal cartilages over the dull space were unusually prominent; but one part did not protrude more than another. In the course of a few weeks, the pain became still more severe. He was unable to lie down at night; the easiest position being a sitting one, with his chest bent forward, "to take off the pressure of his breastbone on a lump in his chest." He entered the hospital, where he died worn out by the pain and loss of sleep, from being unable to lie down. I received a note stating that the pericardium was cancerous. In the other case the upper surface of the liver, the diaphragm and the under surface of the right and left lung, were affected with cancerous tubercles. The pericardium and the surface of the heart were also studded with them, but not to any very great extent. The lining of the pericardium and of the surface of the heart was congested, and the pericardium contained from eight to ten ounces of serum. The symptoms were throughout those of cancer of the liver. The patient was 45 years of age. He had drank heavily, and been exposed to great hardships and mental trouble.

Primary cancer seems to be very rare. In a case recorded by Dr. Paiket,* paralysis of one side of the body and face occurred in the July of the previous year, he became suddenly unconscious, and after recovery, was found to be again paralysed, the tongue being also affected. There was a murmur heard over the whole of the heart, instead of both sounds. The same sound was heard in the large vessels. He improved very much up to the 15th of July, when an attack of acute palpitation occurred, followed by complete paralysis;

* Rankin's Retrospect, 1868, cited from Schmidt's Jahrbücher, 1865.

the next day the pulse fell from 126 to 96. Similar attacks occurred on the 26th and on the 3rd of August, when the pulse disappeared from the radial and ulna arteries. On the 4th, the pulse returned. He died apoplectic the same evening.

The pericardium contained half an ounce of serum. In the left ventricular septum, near the ostium arteriosus, there was a soft nodulated tumour, the size of a chestnut. It protruded into the right ventricle. The aortic valves were involved in the tumour. There was an embolus in the brachial artery at its division into the radial and ulna, like the tumour in the heart.

It is to be regretted that the state of the cerebral arteries was not noticed.

5. HYDATIDS.—The heart, like any other organ, is liable to become the seat of hydatids, and this may take place either in a *primary*, or in a *secondary* form ; the cysts extending from the left lobe of the liver through the diaphragm, or from the lungs to the pericardium.

(1.) *As primary.* This seems to be of rare occurrence. I have a preparation in my possession, taken from a patient who died from phthisis, in which there is a cyst rather larger than a small almond in the interventricular septum, and which protrudes slightly into both ventricles. The patient, a female, aged 20, had never complained of her heart. Dr. Crook has kindly furnished me with the notes of the case of a lad, aged 16, a ropemaker, previously in good health, who while walking home from his work, fell down complaining of giddiness, and died almost immediately. There was a cyst in the walls of the left ventricle of the size of a pigeon's egg. The ventricle was empty.

In the colony, rarely more than one cyst is met with, but the case in which an hydatid existed in the right ventricle, with free hydatids in the right pulmonary artery, is peculiar.* The patient, aged 23, single, entered King's College Hospital, under Dr. Budd, the 23rd of December. She had had two attacks of pleurisy, and was suffering on admission with symptoms of congestion of the lungs, with expectoration of mucus, tinged with blood. A faint systolic bellows murmur was heard. On the 12th of April, she complained of severe pain shooting through the left side of the chest, and from this time she complained of pain in the præcordia. There was extensive dulness on percussion in the præcordial space, but no unnatural bellows murmur. There were indications of bronchitis on the upper part of the left lung in front, and in both lungs posteriorly. There was great difficulty of breathing at times, and the face, which throughout had a purplish tint, was expressive of great distress. Dropsy of the legs and abdomen set in. On the 4th of May, she was found pale, gasping at long intervals, and pulse imperceptible. She died five minutes afterwards.

There was old pleuritic adhesion. The pericardium contained an ounce of serum, and posteriorly it was adherent to the heart. The heart was irregularly shaped. In the apex of the right ventricle, there was an hydatid about the size of an orange; it projected into the ventricle. This ventricle and the right auricle were distended with clotted blood. The left side of the heart was empty. There was a small free hydatid under one of the pulmonary valves; and a second one immediately above, and several others just before the artery divided. In the left pulmonary artery, and chiefly in the upper lobe of the lung, there were several clusters of hydatids and the empty skins of hydatids. The hydatid tumour in the heart

* Medical Gazette, 1858,

was stuffed full of hydatids, and from it the hydatids in the pulmonary artery had escaped.

6. ANGINA PECTORIS.—Only 3 of the 310 cases which presented themselves at the institution were suffering from angina pectoris. In each of the cases there were indications of ossification of the aorta and of the smaller vessels of the heart and pericardium. The arcus seniles was strongly marked in each case. Age, however, accounted for this, one being 60, one 67, and the third, 73 years of age. In each case, the pain occurred on exertion, ceasing on resting for several minutes; sometimes returning, sometimes not, on again walking. One case died from rupture of the heart. It may be, I think, expected that this disease will become more common than it is at present.

CHAPTER X.

ANEURISM OF THE THORACIC AORTA.

ANEURISM of the aorta is much more common in the colony than in England; and although the mortality from other diseases is diminishing, it seems to be increasing, particularly in the mining districts.

In Melbourne, in one year, there were 26 deaths from aneurism out of 3593 from all causes—about 1 in 138; while in London, only 68 died out of 59,103—about 1 in 869. In England, the proportion of deaths was less than in London; for in one year it amounted to only 321 out of 419,865 deaths from all causes. In Glasgow, in 1865, there were only 8 deaths registered from aneurism in 13,912 from all causes; and in 1866, only 5 in 12,896. In Edinburgh, in 1865, the number of deaths registered was higher than in Glasgow, being 21 in 4853 from all causes; but in 1866 the number was 14 in 4811. In Glasgow, in 1857 and 1858, the deaths were more numerous, being 13 in the former year and 15 in the latter; and less in Edinburgh, being 7 in 1857, and 10 in 1858. The population of Glasgow, in 1861, was 329,097, and of Edinburgh, 274,093.

In 1868, the deaths in the colony amounted amongst males to 57 out of 5865 from all causes; but amongst females only to 5 out of 4202. (The population was estimated at 671,222.) In Monmouthshire and Wales, having together a population of

1,212,834—nearly twice as numerous as that of this colony—
there were only seven deaths registered—5 males, and 2 females.
The following table will show the number of deaths from
aneurism, in the colony, from 1859 to 1868, the deaths from all
causes, and the deaths in every 1000 of the population. It will
show that although the mortality from aneurism has increased
from 23 in 1859 to 72 in 1866, 73 in 1867, and 62 in 1868, the
population has not increased in a corresponding degree; for in
1859 it was estimated at 517,226, and in 1868 at 671,222. The
number of deaths to every 1000 of the population was, in 1859,
18·31; in 1860, 22·36; but in 1868 it was 15·00. The two previous years—1866, it was 19·37; in 1867, 18·06; but in 1864 it
was 15·08.

	Deaths from Aneurism.		Total.	Deaths from all causes.		Deaths in 1000 of Population.*
	Males.	Females.		Males.	Females.	
1859	22	1	23	5721	3748	18·31
1860	31	3	34	7134	4927	22·36
1861	51	3	54	6124	4398	19·45
1862	48	1	49	5900	4180	18·39
1863	45	2	47	5646	5856	16·91
1864	63	3	66	5202	3685	15·08
1865	49	5	54	6158	4303	16·97
1866	67	5	72	7016	5270	19·37
1867	63	10	73	6613	5120	18·06
1868	57	5	62	5865	4202	15·00

ADMISSIONS AND DEATHS IN PUBLIC INSTITUTIONS.

The Melbourne Hospital reports are of comparatively little
value prior to 1863. They show, however, a very great increase
in the number of admissions and the deaths from this disease. In
1857, with 1683 cases treated, 8 cases of aneurism were admitted—

* Extracted from vital statistics of Victoria,

3 died; in 1858, with 2013 cases treated, only 3 were admitted— 3 died; in 1859, there was only 1 case of disease of arteries; in 1860, 17 cases of disease of arteries, in 3628 cases treated; in 1861, in 4309 cases, there was only 1 case of disease of artery; in 1862, 3179 cases were treated, 9 cases of aneurism were admitted, of which number 3 died, and 10 of disease of arteries, of which 9 died. In

1863—3147 cases were treated,	20 of aneurism,	2 died.
—64—3024 ,, ,,	14 ,,	7 ,,
—65—3253 ,, ,,	19 ,,	9 ,,
—66—3300 ,, ,,	16 ,,	6 ,,
—67—3095 ,, ,,	16 ,,	6 ,,
—68—3255 ,, ,,	14 ,,	10 ,,
—69—3355 ,, ,,	26 ,,	8 ,,

A very large number of the cases of aneurism treated in the Melbourne Hospital come either from the country or from the adjoining colonies—particularly New Zealand. Of 41 cases of aneurism of different arteries which I have collected, only 12 were resident in Melbourne and the suburbs. From its being more frequently met with amongst diggers and men employed in carrying heavy weights on their shoulders, as bricklayer's labourers, &c., and sailors who are compelled to use sudden exertion, fewer cases will be found to be admitted into the hospitals situated in the agricultural than in the mining districts. I have been informed that when flogging was much used in the prisons here, and in Tasmania, aneurism was frequent. As many as 10 cases were admitted in one year into the Ballarat Hospital, although the number of all cases treated only amounted to 1090. In the Beechworth Hospital there were 3 cases in 392 admitted. But the number admitted into the Pleasant Creek Hospital was still larger, being 5 out of 360. This is a much higher rate than

even in the Melbourne Hospital, where there were only 14 cases
in 3255 cases. The following table will show the number of cases
admitted, deaths, and number of cases of aneurism and deaths in
the Victorian Hospitals :—

	No. of cases treated.	Deaths from all causes.	Cases of Aneurism treated.	Deaths from Aneurism.
Amherst	217	24	1	1
Ararat	285	26	—	—
Ballarat	1090	113	10	2
Beechworth	392	50	3	2
Belfast	44	6	—	—
Bendigo	734	89	2	1
Castlemaine	620	46	1	1
Creswick	238	17	—	—
Daylesford*	150	13	2	2
Dunolly	244	28	3	—
Geelong*	958	64	4	2
Hamilton*	210	17	—	—
Heathcote	92	11	1	—
Inglewood	123	12	1	1
Kilmore	125	9	—	—
Kyneton	289	19	—	—
Maldon	91	12	—	—
Maryborough	274	20	1	1
Melbourne	3255	423	14	10
Pleasant Creek	· 360	27	5	—
Portland*	27	2	—	—
Sale	136	12	—	—
Swan Hill	108	3	—	—
Warrnambool	91	5	—	—
Wood's Point	99	4	—	—

A page might be filled with the different opinions given, even
in one case of aneurism, from the absence of the bellows sound.
In a near relation of my own, its absence led a medical man to
propose to plunge a lancet into it to let out the matter.

The bellows sound—upon the existence of which so much stress
is usually laid—did not occur in more than six of the twenty cases
in which the aneurism was situated in the right half, or ascending
portion of the vessel, and in only two of the six cases in which it

* These are also Benevolent Asylums.

was seated in the descending portion.   In one of the first two cases it was only observed at the commencement of the disease; and in the other, after the aneurism had penetrated through the cartilages of the ribs.   It was not heard in this case in the front of the chest; but along the right side of the spine, for a distance of four fingers' breadth, it commenced at a point corresponding to the lower part of the aneurism in front.   The tumour in front measured nine fingers' breadth in length; and in shape resembled half of a small cocoanut.

In one of the cases in which the sound existed in aneurism of the left side of the arch, the subclavian artery was given off from the sac.   It was only heard along the course of the subclavian artery.

In one of the cases in which there was an aneurism just at the commencement of the aorta, there was a loud murmur heard over the aortic valves.   After death, these valves were found covered with fibrinous growths.   In two of the other three cases, in which the aneurism was seated in this part, a feeble bellows sound was heard in the pulmonary artery, but it was apparently caused by pressure on this vessel.

The following diagram will show the position of the aorta from its commencement at the base of the heart to the lower border of the root of the left lung :—

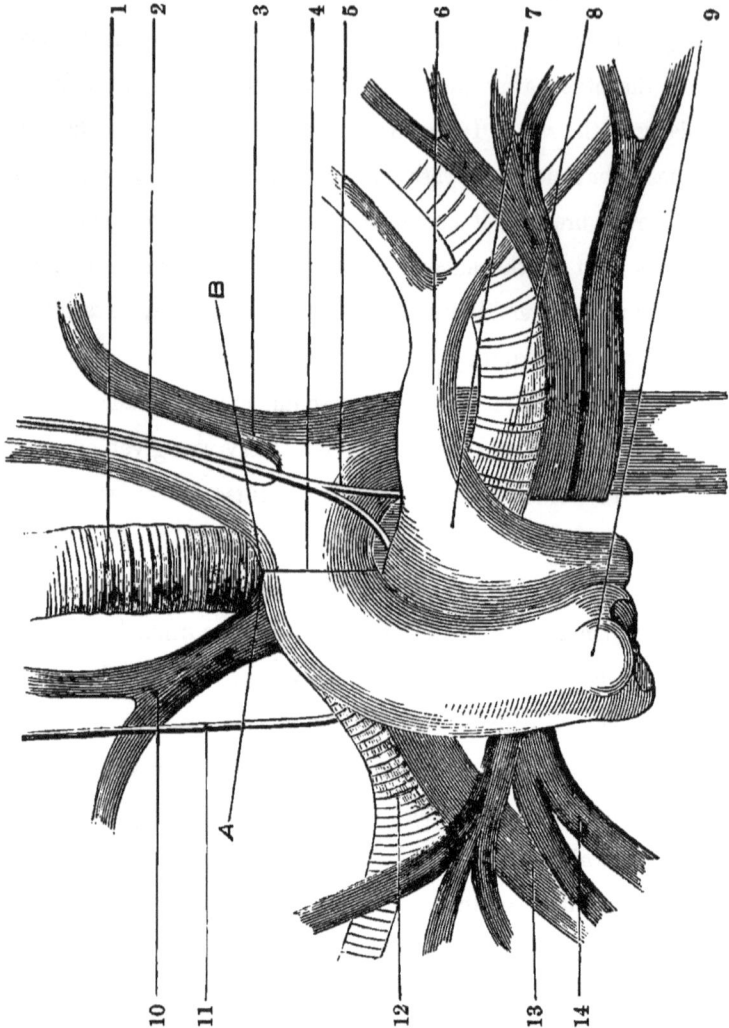

A Right half of the Thoracic Aorta. | B Left half.

1—The trachea dividing behind the arch of the aorta into the right and left bronchus; the left bronchus (8) passing under the arch and in front of the descending aorta.

2—The left carotid artery, given off from the arch of the aorta.

3—The left subclavian artery, also given off from the arch.

4—The arch of the aorta.

5—The left pneumogastric nerve, giving off as it passes in front of the arch, the recurrent laryngeal nerve, which ascends behind the arch.

6—The left branch of the pulmonary artery.

7—The trunk of pulmonary artery dividing into left and right pulmonary arteries; the left one (6) lies in front of the left bronchial tube (8) and the descending aorta.

8—The left bronchus, with the left pulmonary veins immediately below.

9—The commencement of the aorta, showing the dilatations which correspond with the three aortic valves. This part of the vessel is very liable to aneurism.

10—The innominate artery, dividing into the right carotid and right subclavian arteries.

11—The right pneumogastric nerve.

12—The right bronchus.

13—The right pulmonary artery.

14—The right pulmonary veins.

A careful consideration of the course and the relations of the aorta from the point that it is given off from the left ventricle to the lower border of the left lung, will be of great assistance in enabling the practitioner to arrive at a correct diagnosis as to the exact position of the aneurism.

The ascending portion of the aorta—see diagram—(A) in its course from the left ventricle lies in *front* of the root of the right lung, while the descending portion (B), lies *behind* the root of the left one. The sternum, and the cartilages of the ribs cover the ascending portion of the vessel, while the descending lies between the root of the left lung and the spine. The vessel in its ascent rests successively on the trunk of the pulmonary artery (7), the right pulmonary veins (14), right pulmonary artery (13), right bronchial tube (12), and then in front of the trachea (1.) In its descent, it lies successively behind the left pneumogastic nerve (5), the left pulmonary artery (6), the left bronchial tube (8), and then the left pulmonary veins.

The part of the Aorta most liable to Aneurism.

Some parts of the vessel are more liable to give way than others; the right, or ascending half (A), more than the left, or descending (B). There are two weak points in the former, namely, immediately above the aortic valves (9), and just at the curve of the arch, in the vicinity of the right bronchial tube (12.) In the latter, the weak point is the vicinity of the left subclavian artery (3).

From the ascending portion of the vessel being exposed to more pressure than the descending, aneurism is more liable to occur in it. Out of 28 cases which I have been able to collect, 22 were

N

situated in this portion of the vessel ; in 7, immediately above the aortic valves, and in 15, near the arch ; while only in 6 cases it was situated at or below the arch, on the left side.

SYMPTOMS.—When an aneurism occurs in either of the three situations, it presents certain peculiarities—

1st. *In the region of the Aortic Valves.*—From the vessel giving way on its deep or left aspect (9), the symptoms are very obscure. In the seven cases in which the aneurism was seated here, there was pain, confined to a small space, close to the aortic valves. It could generally be traced to over exertion, a fall, or being kicked or jumped upon, or carrying a heavy load on the right shoulder. The pain generally occurred suddenly, any exertion aggravating it, the pain remaining constant, or not lessening in intensity, for several days, and then very slight exertion brings it back with more or less intensity. From the aneurism being small, and communicating with the aorta by a minute opening, there is neither tumour nor bellows sound to be discovered. In the four of the seven cases in which death ensued, the aneurism varied in size from a small to a large walnut Three were filled with fibrinous layers. In one of the number death ensued from rupture of the aneurism, the blood escaping into the pericardium. See *Rupture of the Heart.*

The existence of an aneurism was also unsuspected during life in the other three cases. In one of them, the patient, a prostitute aged nineteen, died from congestion of the lungs. During life, a loud murmur was heard over the aortic valves; this was found to be due to the presence of dense, warty growths on them. She had more pain in this region than could be accounted for by the growths. In the third case, there was disease of the right side of the heart; in the fourth, there was a loud murmur heard over the pulmonary valves. He had great difficulty in walking, being compelled to stop every twenty or thirty yards to relieve the breathing, and allow the palpitation of the heart to subside. These symptoms depended on the pressure of the aneurism on the pulmonary artery, just above its valves. There was no disease of its valves.

An aneurism in this situation may open into the pulmonary artery, and produce more or less symptoms of cyanosis, according to the size of the opening, and the quantity of venous blood which gets intermixed with the arterial. See *Diseases of the Pulmonary Artery.*

2nd. *At or near the right arch of the Aorta.*—From the aorta lying to the right of the sternum, and only covered by the cartilages of the ribs, the presence of the aneurism is early discovered. With the pulsating tumour there will be more or less pain, always aggravated by exertion, and which, when the ribs begin to erode, will become of a dull, boring character. Occasionally the aneurism forms on the posterior aspect of the vessel, but this is of rare occurrence; it happened in only two of the fifteen cases. In both these cases the pain was referred to the region of the scapula and the shoulder. This led to the disease being considered as rheumatism. This might very easily happen when there is no indication of tumour in front, and no bellows sound. With the pain the patient complains of great difficulty, when the aneurism presses on the bronchial tube, in walking far without stopping to take one one more deep inspirations to fill the chest. He is then able to go on again for about the same distance. In these cases, by placing the stethescope over the seat of pain, and requesting the patient to cough, the mucus in the bronchial tube will be heard to strike against some obstruction. This is increased by pressing with the finger over the spot, and by throwing back the head and chest. In a case which has lately come under my notice, in which there was an exudation of bloody fluid into the right bronchial tube, this sound was very strongly marked. The indications of aneurism externally were not very evident.

There is another symptom sometimes met with, namely dilatation of the external jugular and the other veins which open into the descending vena cava. In one case in which the innominate artery was implicated, the right jugular vein was dilated, but usually when the aneurism becomes large, the descending vena cava is pressed upon, the external jugular veins, and the superficial veins of the arms and chest then dilate.

In the early stage, the aneurismal tumour gives a thrill to the finger at each pulsation, but later, as the walls thicken and become less elastic, it is replaced by a "thud"-like impulse.

3rd. *From the centre of the arch to below the root of the left lung*.—This part of the vessel is much less liable than the ascending portion of the vessel, from its not having to bear the weight of the ascending column of blood. Of 28 cases, aneurism occurred in only six of the number in this situation. The centre of the arch is rarely the seat of aneurism. It may either press on the bifurcation of the trachea, or one or other of the bronchial tubes, or on the œsophagus.

In a case in which the sac was situated at the posterior part of the arch, the patient, a stonemason aged 45 years, had great difficulty in swallowing, from some obstruction in the œsophagus. During the last three months of life, the food was returned in a semi-digested condition, as if it had been lodged in a pouch, from three to six hours after it had been taken. He experienced great difficulty in swallowing any fluid. If more than two teaspoonfuls were taken at a time, it entered the stomach very slowly, he said ; but if a larger quantity, it was returned in a short time mixed with mucus. He was nourished by beef-tea injections, rubbing in lard over the body, and sipping, as often as possible, without distending the sac in the gullet, strong beef-tea, chicken broth, and water mixed with brandy. By this plan, his strength was kept up for two months. At the end of this time the fluid, when returned, was occasionally tinged with blood. This tinging was thought to be due to some fungous growth. His aspect was healthy, and he only complained of a feeling of weight, with difficulty of breathing, on walking.

The most careful examination of the chest only led to the detection of a little mucus râle in the course of the right bronchial tube, as far as it divided into its smaller divisions. A râle also existed in the left tube, louder than that in the right; but it did not extend beyond the first division of this tube. When the mucus accumulated to a certain extent, cough was excited, and continued until it was expectorated.

At the time the fluid—returned from the gullet—began to be tinged with blood, that from the bronchial tubes became also tinged. It was thought by a medical man, who saw him subsequently to the writer, that no harm could follow the introduction of an œsophageal tube. This was done as gently as possible, but without being able to overcome the obstruction. Pain was excited, and the fluids from the œsophagus, and the mucus from the bronchial tubes, became more tinged with the colouring matter of the blood. He died suddenly on the night of the third day after the introduction of the tube. After death, a small aneurism was found at the posterior part of the arch of the aorta. It had pressed on the œsophagus, and had opened into it. The trachea, at its point of bifurcation, was closely connected with it.

The only circumstances which could lead to the suspicion that an aneurism existed were the symptoms appearing after an effort made to raise a heavy weight—the fixing of the chest causing the vessel to give way, and the exudation of the colouring matter of the blood through the walls of the aneurism tinging the fluid in the œsophagus and the mucus in the bronchial tubes. This tinging, when observed, must be considered of great value in aiding the diagnosis of aneurism in this region.

The occurrence of an aneurism near the pneumogastric nerve may produce symptoms of laryngeal disease. The larynx has been opened several times, but without any permanent benefit. The division of the recurrent laryngeal nerve might be attended with more benefit. In the cases that I am acquainted with, the aneurism was not discovered till after death.

When the aneurism occurs near the origin of the left subclavian artery, and left carotid arteries, a bellows sound will be heard in the course of these vessels, if they are given off from the aneurism. In the case of a brickmaker aged 35 years, strong and healthy-looking, the symptoms of aneurism were so very obscure that he was twice turned out of the hospital as a malingerer. The disease was excited by carrying heavy logs of wood on his left shoulder; at first he complained of pain in the left nipple. It was always aggravated permanently, by over exertion. When seen he had not been able to lie down for nine days.

He was sitting up in bed, leaning towards the left side, with his arms on a box, and his head sunk deeply between his shoulders; pulse 90 in both wrists, but feebler in the left. Near the upper part of the left side of the sternum, a feeble pulsation could be discovered, with some dulness on percussion, and absence of respiratory sound; but there was no increased vocal resonance. There was a distinct bellows sound heard over the left carotid and left subclavian arteries. He died suddenly four months from the time he was first seen. There was an oblong aneurism extending from the right of the left carotid artery to below the left subclavian; both these vessels were given off from the sac. It had opened into the left bronchial tube. The vertebral column was eroded. The walls of the sac were thick, but there were no fibrinous layers in it. There was no ætheromatous deposit in any of the large arteries of the chest.

In another case, in which deep-seated pain existed between the second and third intercostal spaces, there was a marked bellows sound heard over the subclavian artery. He was a butcher, and attributed the pain to carrying heavy loads on the left shoulder. He was left-handed, and in using the saw, said it caused a "jerking pain" in the aneurism.

In a case now under the care of Dr. Gregory and myself, there was when first seen no pulsation in the left wrist or brachial artery, the muscles of the left arm were softer and much smaller than those of the right. There was an indistinct pulsatory tumour to the left of the sternum in the second intercostal space. There was a distinct bellows sound in the tumour, which could be traced both up the left carotid and left subclavian. He complained more of pain in the back along the intercostal spaces, and under the scapula, than in the aneurism. By strict rest and the exhibition of iodide of potash and digitalis, the pulsation in the left wrist returned. The tumour lessened in size, and the pains in the side disappeared. Of late, the perchloride of iron has been given with digitalis, with marked benefit. The potash reduced his strength too much. He had been a storeman. He was delicate looking and thin. I have seen another case in which the aneurism was situated lower down the vessel near the left bronchial tube. The patient was a clerk in a merchant's office, and was about 40 years of age. He had deep-seated pain in the left side of the chest, above the nipple, near the third intercostal space, passing through to the shoulder blade. Walking fast or riding in a cab over a rough road, greatly aggravated the pain. In the morning as soon as he sat up in bed, he was seized with a peculiar long harsh cough, not unlike hooping cough, which continued until vomiting set in.

Disease of the pulmonary valves is very apt to resemble aneurism of the lower part of the left side of the arch. There is pain, and the same difficulty in walking far without resting, as in aneurism. The pain is, however, at the base of the heart, and the valvular sounds are either soft or rasping. I have not seen a case of aneurism of the pulmonary artery in the colony. The effect of the pressure of the aneurism at the commencement of the aorta on this vessel has been already alluded to. I have seen a preparation in which the sac opened into this vessel. This is a result which may be expected in aneurism of the lower part of the aorta.

Influence of Treatment.—Iodide of potash, with digitalis and nitrate or carbonate of potash, seem to give more permanent relief than any other medicines. When the pain is severe great relief is experienced from the application of ice. When the patient is pale and emaciated, iron may be substituted for the iodide of potash with benefit. The veratrum viride, from its effect in controlling the action of the heart, is of great service in those cases in which the patients are strong and robust, the pulse full and quick, and the pain in the aneurism severe. It is better, I think, adapted for use in hospitals, where its effect on the circulation can be watched, than in private practice. I have given it occasionally with great benefit, but as a rule—and particularly when the circulation is feeble—its too free exhibition, or continuing it too long, is apt to induce fatal syncope. I was first induced, in 1854, to use the iodide of potash, with digitalis, arsenic, and nitrate of potash, in a case of fibrinous deposit on the aortic valves. While under treatment, a very painful and rather hard pulsating tumour, of the size of a very small orange, was discovered in the superior mesenteric artery.* The fibrinous deposit was removed from the aortic valves, and, at the same time, the tumour in the abdomen became softer, and ceased to be painful. The medicines excited considerable nausea at the end of ten days. The dose was then diminished one-half. He was under treatment for three months. During the whole of this time he was confined to the bed or sofa. The tumour had not increased when seen some months later.

The value of these medicines was very manifest in three cases in which the symptoms of aneurism in the first part of the aorta were well marked. They all did well. This seemed to be due to the aneurismal tumours being small, and of recent formation.

---

* This case is more fully reported in the third volume of the " Medical Record," page 65.

In the case of a Mr. Grocott, a confectioner, the tumour—which was large, hard, and protruding through the right inter-cartilaginous spaces, the veins of the neck and arm being considerably enlarged from the pressure on the descending vena cava—became very much smaller and softer, and the sound lost the peculiar thud observed in old aneurismal tumours, from the absorption of the fibrinous layers, and assumed more the thrill of a recent one—the veins of the neck and arm diminished very much in size. The portion of the sac which was protruding through the inter-cartilaginous space disappeared, and he could lie on his back, with his shoulders a little raised, from there being then less pressure on the right bronchial tube. He was able to walk about and follow his occupation for some time. From indulging in drink and over-exertion, the sac increased in size, but it spread to the right and left, and not externally. He now began to spit up some mucus tinged with red, showing that transudation of the colouring matter of the blood was taking place. He could only lie with ease on the left side, or by bending the chest forward. The right external jugular vein became as large as the little finger, and the veins of the arm and right side of the chest were also very large. From the tumour pressing on both bronchial tubes, very little air entered the lungs. A week or two later, the difficulty in breathing became so great that he was not able to lie down. Neither digitalis, either with or without iodide of potash, nor veratrum viride, produced any effect on the size of the sac. He died somewhat suddenly. Dr. Cutts, who made the *post mortem*, informed me that the sac, which was very large, had not given way.

In the case of the brickmaker just mentioned, the effect of the treatment was well shown. In six days the pain lessened so much that he was able to lie down for a short time, and, at the end of three weeks, for six hours, and the pain gradually disappeared. He lived for four months in a state of comparative comfort, although confined to bed.

I have lately seen a case under the care of Dr. Crooke. The patient (a tailor, of the name of Thomas, residing in Collingwood) received at first much benefit from the veratrum viride, and, of late, from the iodide of potash. From the kindness of Dr. Figg, of Williamstown, I have been able to watch the case of a patient of the name of Norris, residing in Lonsdale-street. Strict rest, and the administration of the perchloride of iron, were of the greatest service in checking the disease in this case. Latterly, the sac made its way through the cartilages of the right side of the sternum. When this occurred, a bellows sound was heard along the right side of the spine for the first time.

In Thomas's case, death ensued from the aneurism encroaching on the bifurcation of the trachia. In Norris's case, Dr. Figg states in reply to a note from me, that "the case you refer to terminated neither by pressure on the lungs nor bursting. The tremendous pulsation caused absorption of the costal cartilages of the right side, and through this aperture; the aneurism passed external to the thorax, having no covering but the integuments and subjacent fascia. At this stage there was very little cough, and no congestion of the lungs; the respiration and pulse were correspondently low; the features livid, with areolæ round eyes and mouth, and the perpendicular position could not be assumed without risk of fainting. I believe the death occurred by syncope, the result of the removal of so large a quantity of arterial blood out of the direct current of the circulation into the cavity of the aneurism, which could not have contained less than a pint."

Termination.—A cure when the aneurism is small, is not impossible, but by strict rest, life may be prolonged for months and even years. When death takes place, it is generally from rupture, the blood escaping into the chest, the air tubes, gullet, or pericardium. In these cases, death takes place more or less rapidly, according to the quantity of blood which escapes. A communication may form between the aneurism and the vena cava, or the pulmonary artery. When this occurs, life may be prolonged for several weeks. When the aneurism gradually extends to the bifurcation of the trachea, death takes place slowly. The same occurs when the spine becomes eroded, and the spinal cord pressed upon.

J. AND A. M'KINLEY, PRINTERS, QUEEN STREET, MELBOURNE.